Rano Musaeva
Nigina Musaeva
Saidzhon Uvaizov

Improving the process of evaporating liquorice root extract

AF723130

Rano Musaeva
Nigina Musaeva
Saidzhon Uvaizov

Improving the process of evaporating liquorice root extract

Monograph

ScienciaScripts

Cover image: www.ingimage.com

This book is a translation from the original published under ISBN 978-620-9-34485-5.

Publisher:
Sciencia Scripts
is a trademark of
Dodo Books Indian Ocean Ltd. and OmniScriptum S.R.L publishing group

120 High Road, East Finchley, London, N2 9ED, United Kingdom
Str. Armeneasca 28/1, office 1, Chisinau MD-2012, Republic of Moldova, Europe
Managing Directors: Ieva Konstantinova, Victoria Ursu
info@omniscriptum.com

Printed at: see last page
ISBN: 978-620-9-78849-9

Contents

This monograph describes the characteristics of the liquorice root evaporation process and, based on an analysis of thermophysical properties, determines the conditions for foam-free concentration of liquorice root extract.

In connection with the intensive development of the agro-industrial complex, the question of the most rational use of natural plant resources for the needs of the national economy has become acute.

Liquorice root is widely used in East Asian medicine. Liquorice is a perennial herb with a fairly robust root system that extends deep into the ground.

In many countries, liquorice occupies a special place among plant-based raw materials. Liquorice root has long been widely used in medicine, industry and agriculture, and is also exported as a medicinal and technical raw material.

INTRODUCTION

According to data from the Chamber of Commerce and Industry of Uzbekistan, over thirty enterprises and companies of various ownership structures are engaged in the harvesting and processing of liquorice root in various regions of Uzbekistan. They are primarily focused on the export market. This is, of course, supported by the fact that liquorice root is experiencing a renaissance in many countries around the world: it is 'consumed' in vast quantities by the global food industry, as well as by pharmaceutical and cosmetics companies. Over 70 per cent of the liquorice grown and processed in our country is exported abroad.

But this is not just about commerce. The cultivation of liquorice in Uzbekistan has an exceptionally important environmental impact. It is a red halophyte. The plant thrives in salty conditions. And where it is grown, the salinisation of the land is reversed.

Traditionally, liquorice root is grown most extensively (due to climatic conditions) in north-western Uzbekistan – in the Republic of Karakalpakstan and the Khorezm Region. Liquorice is known as 'boyan' in Karakalpakstan. It grows in the Amu Darya valley and delta, in the tugays, along irrigation ditches, canals and drainage channels, on salt marshes and fallow land. The flora of Karakalpakstan includes two species of liquorice – the naked liquorice and the Zaisan liquorice.

In Uzbekistan, there are a number of enterprises engaged in the cultivation and processing of liquorice root. OJSC 'Karakalpakboyan' (Karakalpakstan, Khojeyli). Its main activity is the harvesting and pressing of liquorice root. In addition, the company is involved in the cultivation of this plant and the growing of liquorice. It was established in 2003 with the participation of an Iranian partner. Over the ten months of last year, the company exported 1,700 tonnes of liquorice root. The countries to which Karakalpakboyan OJSC consistently exports its liquorice root are the USA, France, South Korea, China, Japan and Kazakhstan.

Lakritsa LLC was founded in 1991 in Tashkent. It is one of the republic's first producers specialising in the processing of liquorice root. The company specialises in the harvesting of liquorice root and the production of extract. The range and production of finished products is quite extensive: thick root extract, sliced root in petal-shaped pieces (of various sizes) and calibrated root sticks (of various sizes). Over sixty per cent of finished products are exported to countries in South-East Asia, as well as to Germany, Hungary and Austria. The main customers are primarily pharmaceutical and cosmetics companies.

Herbex Refine LLC is the largest in the Tashkent region. Its processing capacity is 60 tonnes of liquorice root per month. The company was established in 2006. The main plantations used by the company are located in the foothills of the Chimgan Mountains in the Yukorichirchik district.

Beer, chewing gum and electrolysis. The substances contained in liquorice roots promote vigorous foaming, which has found widespread application in cosmetic procedures. Foaming liquorice baths have a beneficial effect on the skin, refreshing it and restoring its natural protective barrier, which is damaged daily by environmental pollution.

In the food industry, liquorice extract is added during the preparation of soft drinks and tonics, to help whip egg whites, as well as to cocoa, coffee, marinades, fruit jellies, and flour-based and sweet products.

Furthermore, liquorice root is used in the production of fizzy drinks, confectionery, as well as in the manufacture of ink, dyes, paper tableware and other products, across a total of twenty industrial sectors. Products derived from liquorice root are widely used in the food industry in baking, and for making sweets, halva and cakes. The extract is widely used in the pharmaceutical industry to produce and improve the taste of medicines, as well as in the tobacco industry and agriculture.

Food industry. Liquorice has long been used in a large number of recipes in sauces and various seasonings as a flavouring agent. Today, liquorice is used in the manufacture of pastes, jellies and similar products as a thickening agent.

The tannins and organic acids ( ) contained in the roots are capable of preserving the freshness of products for a long time, which is utilised in the production of bakery goods. Liquorice also enhances the flavour of the product and prevents oxidative processes.

Confectionery industry. Glycyrrhizic acid, found in liquorice, is widely used as a sugar substitute in the production of lollipops, dragees, sweets, chocolate products, chewing gum, toffee and other confectionery items. In the production of halva, liquorice is used as a foaming agent, giving the product a light and crumbly texture. Today, virtually all sweets contain liquorice root extract, which demonstrates the great popularity and beneficial qualities of this product.

Production of low-alcohol and non-alcoholic beverages. Liquorice is used as a sweetener and foaming agent. Dark and light beers, kvass, syrups, lemonades and other drinks cannot be produced without the addition of liquorice root. A clear example is soft drinks such as Coca-Cola and Pepsi-Cola, which invariably contain liquorice. It is the most popular sugar substitute used in industrial production today. From one tonne of liquorice root, 150–200 kg of sweetener with a sucrose equivalent of '50' can be obtained.

Perfume and cosmetics industry. Compounds of glycyrrhizic acid and its monoammonium salt (glycyrrhizin) possess pronounced anti-inflammatory properties, promote rapid regeneration of skin cells, improve blood supply and alleviate allergic reactions.

Liquorice extracts are used in various types of cosmetic products, including therapeutic ones. Lotions and shampoos that restore hair structure and prevent dandruff, toners and body lotions are one area of the cosmetics industry that utilises therapeutic liquorice root. Other areas include body care products: creams, toners, lotions and cleansing cosmetics.

The tobacco industry. Liquorice root extract has been used for many years in the manufacture of cigarettes as an agent that improves the taste of tobacco smoke and reduces the sensation of a dry throat. Liquorice is also used to sweeten and harmonise the bouquet of tobacco smoke. Few people realise that popular 'American-style' tobacco blends cannot do without the addition of liquorice root to their composition.

Agriculture. Liquorice root is used as a well-established feed supplement in livestock diets. Liquorice leaves contain 5.8 per cent fat (on a dry weight basis), 6–10 per cent protein, and the protein content is 10–15 per cent higher than in alfalfa. Due to the presence of oestrogens, liquorice roots act as a stimulant for meat and milk production. It is also used to feed bees.

Industry. The ability of liquorice extracts to form foam is widely used in metallurgy, PARTICULARLY IN the electrolysis of non-ferrous metals. On the surface of electrolysis baths, toxic vapours of sulphuric acid and zinc sulphate are completely precipitated in the liquorice foam. This accessible and inexpensive method reliably protects workers' health and the microclimate within dozens of production workshops, thereby preventing the onset of a number of occupational diseases. The wide range of applications for liquorice root extract in the chemical industry (as a foaming agent in fire extinguishers and in the manufacture of art ink, waxes, gutalin and dyes) further enhances its practical value.

At present, certain processes for processing liquorice root have been studied and equipment for carrying out these processes has been developed. The process of concentrating liquorice root extract has been studied to a much lesser extent. Many years of operation at plants using intermittently operated open-type evaporators, employed to produce a concentrated solution, have revealed a number of shortcomings inherent in these designs.

The introduction of high-capacity evaporators into production makes it possible to reduce material and energy costs for the evaporation process and improve the quality of the resulting extract by operating the process under optimal conditions.

With a view to the rational use of natural resources, increasing the volumes of production and industrial processing of liquorice by attracting investment from domestic and foreign companies to organise the production of export-oriented products with high added value, as well as fundamentally improving the management system in the sphere of production and industrial processing of medicinal plants, the Cabinet of Ministers has resolved:

1. To approve the proposal of the Ministry of Agriculture and Water Resources, the Ministry of Economy of the Republic of Uzbekistan, the State Committee of the Republic of Uzbekistan for Forestry, the State Committee of the Republic of Uzbekistan for Ecology and Environmental Protection, and the National Bank for Foreign Economic Activity of the Republic of Uzbekistan regarding the establishment of an Association of Organisations for the Production and Processing of Liquorice.

2. The main tasks of the Association shall be:

- coordinating the implementation of programmes for the comprehensive development of liquorice production and processing, and facilitating the formulation of a unified scientific, technical, technological, investment and export policy in this sector;
- the gradual increase in the volume of liquorice root production through the establishment of specialised plantations in regions favourable for its

growth, with the introduction of intensive cultivation technologies and the rational use of natural cultivation areas;

- ensuring the efficient use of liquorice raw materials, organising their deep industrial processing and the production of export-oriented products with high added value;
- participating in the drafting of regulatory and legislative acts aimed at developing the liquorice production and processing sector;
- actively attracting foreign investment and overseas specialists and consultants to the processes of establishing new and developing existing facilities for the cultivation and processing of liquorice root, and the introduction of advanced technologies in this field;
- participation in the effective organisation and coordination of work on the training and retraining of personnel for this sector, and the improvement of their qualifications at higher and secondary specialised and vocational educational institutions, including those abroad.

CHAPTER 1

1. IMPROVING THE EVAPORATION PROCESS LICORICE ROOT EXTRACT

1.1. Characteristics and comprehensive use of

.

A thick liquorice root extract is obtained by concentrating a liquid extract derived from the roots and underground shoots of perennial wild herbaceous liquorice plants (Glycyrrhiza glabra).

Several species of liquorice grow in Central Asia. There are two commercially important species of liquorice: naked liquorice and Ural liquorice. The main stands of naked liquorice are found in Turkmenistan and Uzbekistan in the Amu Darya river basin, whilst Ural liquorice is widespread in the eastern and southern regions of Kazakhstan and Kyrgyzstan, as well as in the southern steppe regions of Western and Eastern Siberia.

Liquorice is a perennial herbaceous plant. As with most typical inhabitants of semi-deserts and steppes, the underground mass far exceeds that of the stems. Over the years, the root system of liquorice occupies large areas; its fast-growing shoots can stretch for tens of metres, sending stems from their buds up to the surface of the earth. In pre-revolutionary times, liquorice thickets were ruthlessly exploited by foreign entrepreneurs as a source of raw materials. In the early 20th century, German firms established themselves in Uralsk, where they built a small factory to process liquorice growing in the Ural River basin. Customs records indicate that in some years, exports of dried root from Russia reached 28,000 tonnes.

Liquorice root is one of the oldest medicinal remedies, which has stood the test of time and is used in modern scientific medicine as a classic agent to improve the taste of various mixtures. Liquorice is used for lung diseases – bronchitis, whooping cough, tuberculosis – as a demulcent and mild laxative. It is successfully used to treat stomach ulcers and is employed as an antidote for drug poisoning, among other uses. Liquorice is one of the few plants whose sweet taste is due to the presence of substances other than sugars. In China, liquorice root is valued on a par with the legendary ginseng.

In East Asian countries, liquorice root and the extract derived from it are ingredients in a great many medicinal preparations. Ancient Arabic medicine held liquorice root in very high regard, which is why it found its way into medical reference books in the 12th century.

It should be noted that the above-ground parts of liquorice are recognised as a highly valuable protein feed. Stems harvested during the flowering stage, as well as the leaves, are comparable in feed quality to alfalfa hay. Waste from extract production (pulp) can be used to make paper pulp (USA), which was used to manufacture disposable glass and earthenware tableware substitutes. In Germany, extracted liquorice root is used as a fertiliser, as it contains up to 1.75% nitrogen and up to 18% minerals.

Apart from medicine, liquorice root extract is widely used in the food industry in the manufacture of certain types of confectionery, chewy sweets and halva, in the production of kvass, sparkling waters and tonic drinks, and in the tobacco industry for flavouring tobacco. In recent years, concentrated extract has been used on a large scale in industries such as non-ferrous metallurgy (lead and zinc production; to suppress sulphuric acid mist in electrolytic baths). The wide range of applications for liquorice root extract in the chemical industry (as a foaming agent in fire extinguishers), the manufacture of artistic ink, waxes, gutalin and dyes also enhances its practical value.

Liquorice root and its concentrated extract have for many years been a traditional export to the CIS, the USA, the UK, Japan, West Germany, East Germany, Hungary and Bulgaria. This is by no means an exhaustive list of the countries that have purchased and continue to purchase liquorice root and its extract from us over the years.

The industrial product—liquorice root extract—is a thick, brown-coloured mass with a faint, characteristic odour and a cloyingly sweet taste with a residual bitterness. When shaken with water, it forms a colloidal, highly foaming solution. Liquorice root extract is valued for its content of glycyrrhizic acid, as well as a number of other associated substances:

Flavonoids – 13.0–4.0%;

Saponins, tannins – 9.7–28.2%;

Extractives – 22.8–44.1%;

Glycyrrhizin – 7.3–23.6%;

Glucose — 0.3–11.0;

Fibre – 9.7–28.2%;

Starch up to 34.0%;

Ascorbic acid – 11.0–31.2 mg%;
Essential oils – 0.03%;
Asparagine – 1.0–4.0;
Lipid-like substances – 0.2–4.7;
Resinous substances – 1.75–4.12;
Proteins – 6.18–10.13;
Gums – 1.5–5.5;
Water-insoluble bitterness -3.0–3.6;
Water-soluble bitterness – 1.8;
Total ash – 4.9–9.7.

A thick liquorice root extract is obtained by extracting the root with hot water, followed by evaporation and concentration of the juice in evaporators.

In terms of quality, the thick extract must meet the following requirements:

Appearance: a thick, homogeneous mass free from lumps and foreign matter;
Colour: dark brown;
Odour: faint, characteristic;
Taste: cloyingly sweet, irritating;
Mass fraction of glycyrrhizic acid: not less than 18%;
Moisture content: 32–38%;
Mass fraction of total ash: not more than 9%;
Mass fraction of substances insoluble in hot water: not more than 2.5%.

The exceptionally sweet roots of the liquorice plant (liquorice root contains glucose and pectin, but its most characteristic component is glycyrrhizin, which has a cloyingly sweet taste: in its pure form, it is approximately 50 times sweeter than sugar) attracted people's attention to it as far back as ancient times. In the oldest Chinese medical texts, liquorice root is classified as a first-class medicinal raw material, approaching ginseng in significance; it appears in them most often under the name 'hongqi' – sweet herb, as well as 'gan-cao' – sweet plant, 'lu-cao' – revitalising plant, and 'mei-cao' – beautiful plant.

The roots of the liquorice plant are mentioned quite frequently in the oldest medical encyclopaedia – the Ebers Papyrus. The ancient Greeks purchased 'sweet root' from the Scythians, who inhabited the vast expanses of the lower reaches of the Danube and the Don. In the 1st century AD, the famous Greek physician Dioscorides named the plant 'glycyrrhiza' for its sweet taste; the ancient Romans subsequently adapted this name to the more phonetically similar 'liquiriza' in their language, which eventually became the shorter 'liquorice'.

The main process for obtaining a concentrated liquorice root extract is two-stage concentration, which is carried out in two stages. The liquid extract obtained in the extractors is evaporated in evaporators or open evaporators to a certain concentration, and the final concentration is carried out in jacket vacuum evaporators. This concentration process is necessary because, with a dry matter content of 35–40% in the extractor, the extract foams heavily in the evaporators and evaporation becomes impossible.

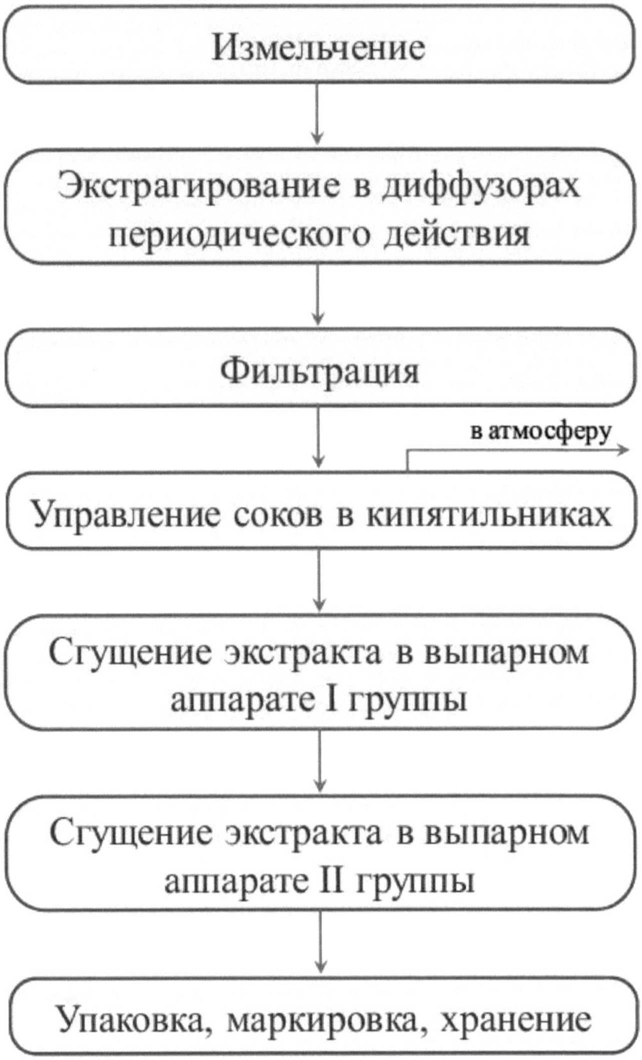

Fig. 1.1. *Process flow diagram for the production of liquorice root extract (method 1).*

Fig. 1.2. *Process flow diagram for the production of liquorice root extract (method 2).*

1.2. Analysis of the operation of evaporators of various designs

One of the most widespread and well-studied technological processes in the chemical, food and related industries is the concentration of organic solutions by evaporation //.

Solutions can be concentrated using several methods: membrane separation, inert gas distillation, ultrafiltration and evaporation.

The evaporation process is one of the key processes in various industries.

The following process parameters have a significant influence on the solution concentration process: concentration time, boiling point and concentration of the product being processed. Consequently, these parameters must be controlled during the concentration process.

The evaporation of solutions from plant-based raw materials is carried out using the energy of a heat transfer fluid under vacuum, at atmospheric pressure, or under excess pressure within the apparatus.

The evaporation process has been studied for many years, and during this period significant progress has been made in developing the theoretical foundations of the process. It can be classified as follows:

1. Evaporation in a layer. This method of evaporation is carried out in a certain volume of liquid accumulated in a specific part of the apparatus, which serves as a sort of reservoir.
2. Evaporation in apparatus with a tubular heating surface and natural circulation of the solution.
3. Film evaporation, characterised by the formation of a liquid film on the heat exchange surface of the apparatus, which may be either rising (ascending) or falling (descending) / /.

The organisation of the evaporation process of extracts from plant raw materials in an ascending liquid film is characterised by a high film flow rate //.

4. The spray evaporation process. In this type of evaporation, a superheated mixture is sprayed into the hollow section of the apparatus using nozzles / /.

All these concentration methods differ from one another in terms of the size of the interface between the liquid and vapour phases. The larger the interface, the greater the rate of solvent removal, which is a key factor in intensifying the evaporation process.

During the concentration process, the quality of the resulting extract depends largely on the process temperature and the duration of exposure to high temperatures. Therefore, the following requirements apply to the process:

1. The process temperature should be as low as possible.
2. The duration of the process must be kept to a minimum.
3. The equipment must ensure maximum removal of the solvent from the

solution at minimum process costs.
4. The consumption of thermal and electrical energy and cooling water by the evaporation plant must be kept to a minimum wherever possible.

Evaporators are classified into the following groups based on their design:
1. Evaporators in which the concentration process takes place in a layer are apparatus with a steam jacket, a coil heating surface and horizontal tubes.
2. Film-type evaporators are mainly direct-flow units with vertical bundles of boiling tubes, operating with an ascending and descending film of the liquid being evaporated.
3. Evaporators with a tubular heating surface and natural circulation of the solution.
4. Spray evaporators.

These evaporators differ from one another in the arrangement and configuration of the heating surface, the type of heat transfer media, the relative arrangement of the working media, the circulation rate and other features. The classification of evaporators is shown in Fig. 3.

The most widely used evaporator designs in the food industry are: those with a steam jacket; vertical evaporators with a tubular heating surface featuring natural circulation and a corresponding superheated space; evaporators with a vertical external tubular heating surface featuring natural circulation; and film-type direct-flow evaporators: evaporators with vertical bundles of boiling tubes, operating with an ascending and descending film of the liquid being evaporated.

Evaporators with a steam jacket. Such evaporators, Fig. 4, are vessels of a specific shape, on the outer surface of which heating jackets (surfaces) are arranged. A heat transfer fluid is fed into the space formed between the outer surface of the apparatus casing 1 and the jacket 2. These apparatuses usually operate intermittently, i.e. a solution of a certain concentration is fed inwards, the heat transfer fluid enters the steam space, and the solution is evaporated to a certain concentration.

The disadvantages of such apparatus include intermittent operation, a low heat transfer coefficient (300–400 *W/m·°C),* low capacity, high metal consumption and intermittent operation.

Vertical evaporator with a tubular heating chamber and natural circulation of the solution. The apparatus is shown in Fig. 1.5. It is more widely used in the food industry. The visible level of the boiling liquid significantly affects the circulation rate and the intensity of heat exchange in these apparatuses. Natural circulation occurs in a system with circulation and boiling tubes.

If the liquid in the boiling tubes *2* is heated to boiling point, then as a result of the evaporation of part of the liquid in the boiling tubes, a vapour-liquid mixture is formed, the density of which is less than that of the liquid itself. Thus, the entire liquid column in circulation pipe *1* is greater than that in the boiling pipes, resulting in orderly flow of the boiling liquid. Natural circulation requires a sufficient liquid level height in the circulation pipes and sufficient vapourisation intensity in the boiling pipes.

A reduction in the level below the optimum causes a sharp decrease in heat transfer coefficients, the formation of emulsion flow regimes and disruption of the washing of the heating surface in the upper part of the boiling tubes, which leads to product decomposition and a significant increase in its colour. At the same time, an increase in scale deposition on the inner heating surface is also observed.

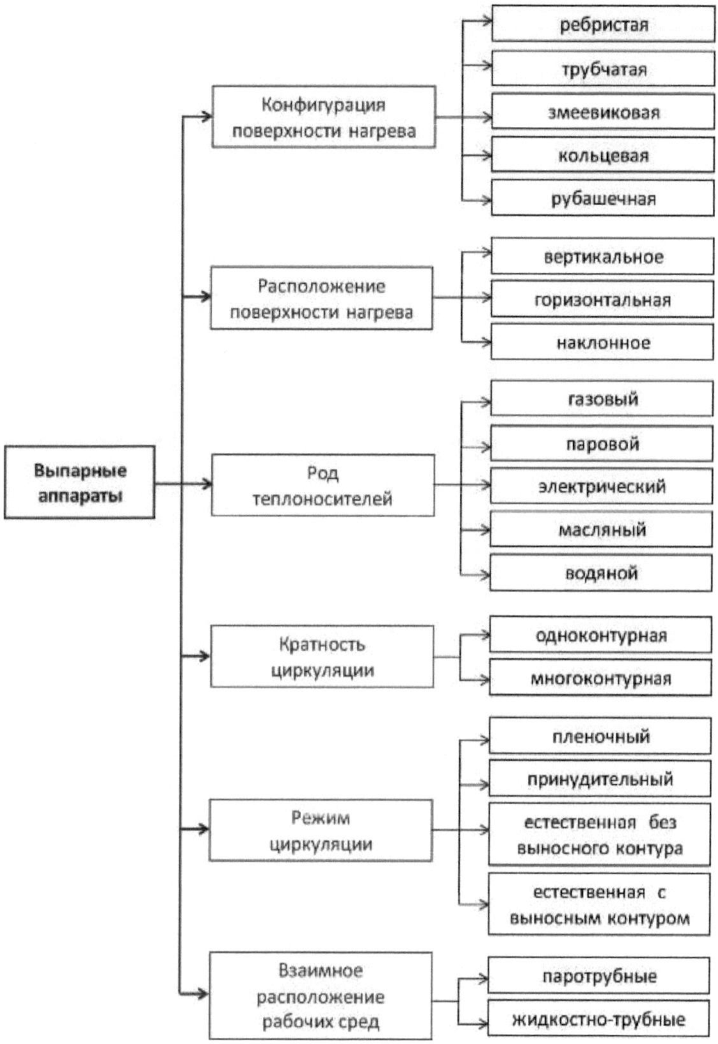

Fig. 1.3. *Classification of evaporators*

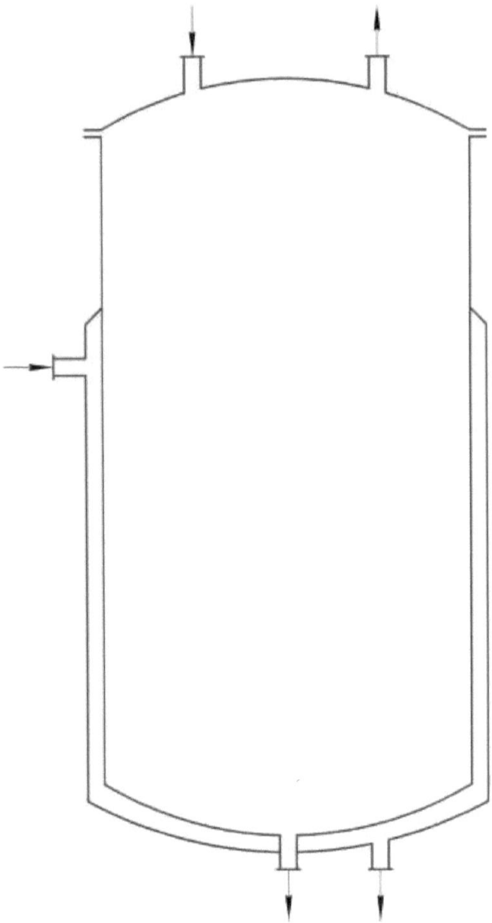

Fig. 1.4. *Evaporator with a steam jacket:*
1 – casing; 2 – heating chamber; 3 – steam space.

From a thermal and technological point of view, the optimal levels of the (sugar) solution in the boiling tubes of the evaporator and the superheater pressure, depending on the solution concentration, are as follows:

Solution concentration, %	Solution level in tubes, %	Vacuum in the superheater space, z, z m/m hour
10	30	1500
25	40	1300

40	50	1100

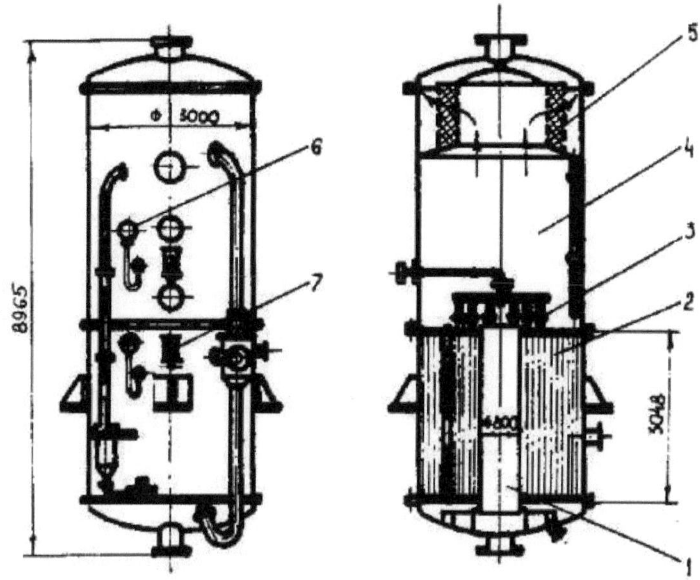

Fig. 1.5 Vertical evaporator with a tubular heating chamber and natural circulation of the solution:
1 – circulation pipe; 2 – heating chamber; 3 – ammonia vents; 4 – headspace; 5 – demister;
6 – pressure gauge; 7 – thermometer.

The heat transfer coefficient decreases significantly as the concentration increases. Thus, at a solution concentration of 10%, the heat transfer coefficient is approximately 2500 *W/m²·°C;* at 25% concentration, it is 910; and at 40%, it is 500.

Apparatus of this design are manufactured with a heating surface area ranging from 500 to 1500 cm^2.

A drawback of the design of these units is the rigid connection of the heating surface's tube bundle to the unit casing, which causes a loss of tightness in the rolled joints due to different thermal expansions of the tubes and the casing.

Apparatus with a remote heating chamber. Such apparatus are shown in Fig. 1.6. By locating the heating chamber outside the apparatus casing, it is possible to increase the evaporation rate not only by increasing the density difference between the liquid and the vapour-liquid mixture in the circulation circuit, but also by increasing the length of the boiling tubes. These apparatuses have boiling tubes, the length of which often reaches 7 *m*. They operate with more intense natural circulation, due to the fact that the circulation pipe is not heated, and the rising and falling sections of the circulation circuit have a considerable height.

The external chamber can be easily detached from the main unit, making cleaning and repairs quicker and easier. The feed solution enters beneath the lower tube grid of the heating chamber and, as it rises through the boiling tubes, is evaporated. Sometimes the feed solution is fed, as shown in the figure, into the circulation pipe. The secondary vapour is separated from the liquid in the separator. The liquid descends through the circulation pipe and mixes with the feed solution, and the circulation cycle repeats. In these apparatuses, a sufficiently intense circulation of the solution is created, as the circulation pipe is located outside the apparatus and is not heated, and the height of the solution column descending down the circulation pipe and the emulsion rising upwards inside the heating tubes is quite significant. The circulation velocity in apparatus with an external heating chamber can reach 1.5 m/s, which allows concentrated and crystallising solutions to be evaporated in them without fear of the heat exchange surface becoming fouled too quickly. Thanks to their versatility, ease of operation and good heat transfer, apparatus of this type have become widely used. A disadvantage of this design is rapid wear, as well as fouling of the upper part of the tube bundle.

Film evaporators. A diagram of a film-type direct-flow evaporator is shown in Fig. 7. The principle of operation of the apparatus is as follows. Juice, heated to boiling point, enters the receiving chamber 7, then into the tubes 6, where it boils and, together with the vapour, moves upwards. After passing through the separator, where the vapour is separated from the juice, the juice then flows through a distribution device into the tubes 5 of the film section of the evaporator and flows down their inner surface in the form of a thin film.

The resulting vapour, together with the concentrated juice, enters the lower separator 9. Secondary vapour is carried via pipes from both separators to the next housing.

Film evaporators fully meet the technological and thermal requirements imposed on evaporators and demonstrate better performance than that achieved in standard evaporators with natural circulation. The residence time of the juice in the thin-film evaporator is significantly shorter, and the unit can operate at a low effective temperature difference. Due to the free swelling of the film of the solution being evaporated, there are no losses of effective temperature difference due to hydrostatic pressure. The small volume of juice present in the unit greatly simplifies automation.

The stable operation of film evaporators depends on the uniformity of the solution distribution across the individual tubes of the heat exchange surface. Figure 6.6 shows diagrams of distribution devices with ring-type water inlets of various designs. Distribution devices A and B are used in foreign-designed units. Analysis of the operation of these distributors shows that these devices do not ensure uniform distribution of the solution across the individual tubes of the apparatus's heat exchange surface. In some cases, the non-uniformity of distribution reaches 20–40%.

The capacity of film evaporators using such spray devices and top-off secondary vapour removal is determined (in *m³/s)* by the equation

$$V_{B} = \mu \cdot f \cdot \sqrt{2 \cdot q \cdot (H - h)} \qquad (1.1)$$

where μ *is* the flow coefficient;

2

j is the area of the strings, *m;*

H is the liquid level on the tube grid, *m; h is* the secondary vapour head, *m.*

9

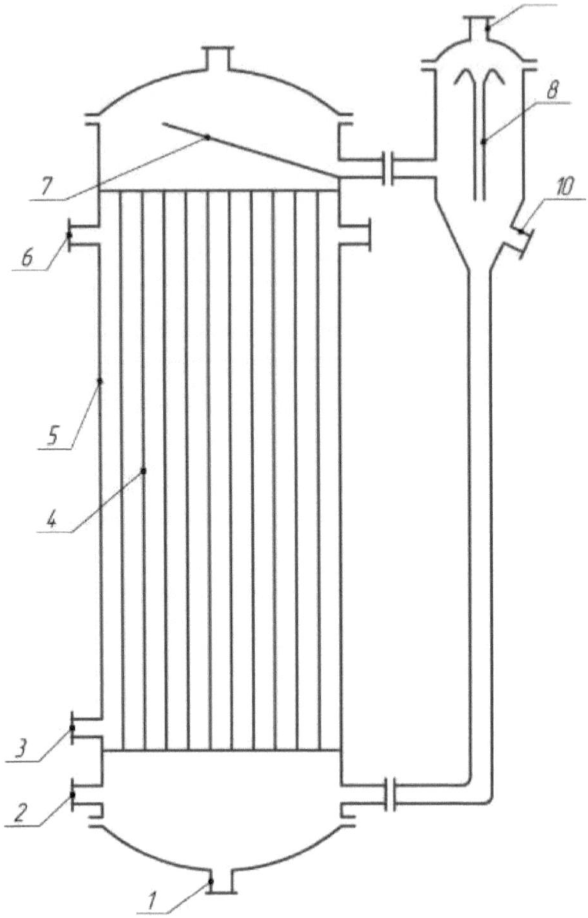

***Fig. 1.6** Evaporator with external heating chamber:*
1 – drain connection; 2 – solution inlet; 3 – condensate outlet; 4 – heating surface; 5 – shell; 6 – steam inlet; 7 – baffle plate; 8 – trap; 9 – secondary steam outlet; 10 – evaporated solution outlet.

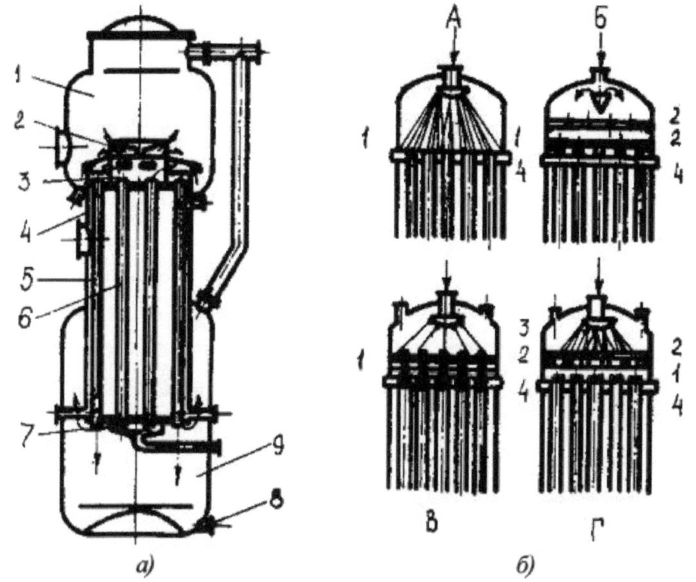

Fig. 1.7a) Schematic diagram of a film evaporator:
1, 9 – separators; 2 – droplet separator; 3 – extension; 4 – casing;
5, 6 – boiling tubes; 7 – receiving chamber; 8 – product outlet;
b) Distribution devices:
1 – annular distributor; 2 – disc; 3 – inserts; 4 – tube grids

In film evaporators, the distribution density usually docs not exceed $0.2 \times 10^{-3}\ m^3/s$. In this case, the height of the solutions above the edge of the evaporation tube is 2.3 *mm.* When the level is reduced by just 0.5 *mm,* the distribution density decreases to $0.14 \times 10^{-3}\ m^3/s$, i.e. by a factor of 1.43.

The distribution devices shown in Fig. 7b (B and G) proved to be more reliable in operation. The distribution of the liquid entering the individual tubes of the evaporator is determined by its overflow through the holes in the distribution disc.

The capacity of film evaporators with distribution devices B and D and top removal of *z*

is determined (in *m/s*) by the formula:

$$V_H = \mu \cdot f \cdot \sqrt{2 \cdot q \cdot H} \qquad (1.2)$$

where: μ - flow coefficient;

2

j – area of the strings, *m;*
H – height of the liquid above the tube grid, *m.*
The flow coefficients μ in equations (1.1) and (1.2) depend on the temperature and density of the solution, and the configuration of the nozzles.

1.3. Foaming phenomena and methods of foam suppression during evaporation

Many processes in the food and other industries are accompanied by intense foaming during their implementation. This is due to the presence in solutions of proteins, amino acids, vitamins, inorganic salts, surfactants and other compounds. The greatest foaming of solutions occurs during their evaporation.

The author of the work / / attributes the causes of foaming to the low surface tension of the solution and the unsatisfactory choice of design and load of the evaporator.

Solutions with high viscosity and low surface tension are prone to foaming, i.e. they form thin, stable films around water vapour bubbles. The resulting foam fills the entire vapour volume and is carried away with the secondary vapour to the next stage of a multi-stage evaporator or to the condenser of a single-stage unit. It is extremely difficult to prevent vapour humidification during evaporation in foam-forming devices. Increasing the size of the steam space does not lead to a significant reduction in steam humidity. Various mechanical separators are also of little effect.

The evaporation process of liquorice root extract is also accompanied by vigorous foaming. In the late 1930s, the Czech scientist Ružička established that liquorice root contains glycyrrhizin. Research has shown that glycyrrhizin consists of the potassium and calcium salts of tribasic glycyrrhizic acid, which is a saponin, i.e. a substance capable of producing abundant foam. When the liquorice root extract is evaporated, foam may enter the steam space of boilers and evaporators, which can disrupt the rhythm of the production process. This results in a significant loss of product with the secondary vapour, a reduction in the evaporator's productivity, a decrease in the heat transfer coefficient, and an increase in the consumption of thermal and electrical energy.

An increase in the concentration of liquorice root extract and a rise in its viscosity increase its foaming capacity: the volume of foam formed when the concentration of liquorice root extract is increased from *10%* w/w to *40%* w/w increases several-fold. The foam of liquorice root extract is very stable.

The evaporation of liquorice root extract in existing industrial evaporators of any design can only be carried out if the operating parameters and evaporation processes ensuring efficient, foam-free evaporation are investigated.

In various industrial sectors, a wide variety of methods are used to combat foaming – chemical, mechanical, combined and others.

Of considerable interest is the preheating of solutions, which reduces foaming during the subsequent evaporation process; the method of foam destruction can be carried out using centrifugal forces in cyclones of various designs, and the method of foam suppression using liquid or gas jets. It should be noted that all existing methods of foam destruction share a common drawback: they incur additional costs, thereby increasing the production cost. ....

1.4. The mechanism of foam formation and its effect on carryover

In the foaming regime, carryover increases significantly. Foam generation and droplet carryover are jointly influenced by the surface energy released during bubble collapse and the kinetic energy of the flow that breaks up the foam and droplets. The number of large droplets and the height to which they are thrown increase. The level of foam formed can reach considerable proportions. In this case, due to a reduction in the height of the vapour space below the critical level, there is a sharp increase in vapour humidity, and entrainment of foam flakes is possible. A further increase in vapour velocity leads to the almost complete disappearance of the foam; however, during this period, the third boiling crisis occurs, accompanied by an increase in vapour humidity and a decrease in liquid flow rate.

V.G. Gleim and B.R. Shidlovsky note in their work that the strength of bubbles entrained from a solution is always higher than that of bubbles obtained from pure liquids.

The dependence of entrainment from solutions of various salts was investigated in the work / /. The authors obtained the following dependence of entrainment on salt content: the entrainment coefficient decreases in the region of low salt concentrations and increases with rising concentration. When the concentration rises above the critical value, the vapour humidity increases sharply. The value of the critical salt content of the solution decreases rapidly with increasing pressure; the value of the critical salt content varies widely depending on the pressure value $\$_{\iota\varphi}$ = 50^ 7000 *kg/kg.*

Foam breakdown and elimination of the foaming operating mode of the apparatus can be achieved by selecting the appropriate hydrodynamic:

operating mode, apparatus design and

the use of devices for mechanical foam suppression, as well as by means of additives that increase the surface tension of the solutions. There are a large number of methods for mechanical and chemical foam suppression; however, the question of their application requires specific research. To reduce foaming during the evaporation of liquorice root extract and a number of other highly foaming solutions, vegetable oil is added to increase the surface tension. The authors of the study provide data on the permissible vapour space pressures based on the evaporated moisture at atmospheric pressure. For water, $A = 2000\ ^{m^3}/m^3 \cdot h$; for solutions, it is half that value; and for foaming solutions, it is a quarter of that value. The height of the vapour space above the surface of highly foaming solutions is taken as $H = 2.5$–3 *m.* Product losses due to carry-over are particularly high when air enters the apparatus at the moment the solution is fed. Foaming is reduced at a certain solution level in the apparatus and with an increase in the regularity of its operation.

Foaming solutions should be fed into the apparatus via a tangentially positioned rectangular transition pipe, elongated vertically. According to the data in the study / /, with foaming liquids it is beneficial to increase the speed and frequency of solution circulation in the evaporator circuit to 20–30 *m/s;* however, this approach is not always feasible or economically viable.

CHAPTER 2

2. THERMOPHYSICAL AND RHEOLOGICAL CHARACTERISTICS OF LIQUORICE ROOT EXTRACT

2.1. Methods for determining the thermophysical properties of solutions

To improve the technological processes involved in the processing of liquorice root, it is necessary to know the properties of the intermediate products, which include liquorice extract. Among the properties of liquorice root extract, thermophysical properties play a significant role, as they are widely used in calculating the parameters of heat treatment.

Thermophysical characteristics (TPCs) – the thermal conductivity, thermal diffusivity, specific heat capacity and density of liquorice root extract – change significantly during heat treatment. In this regard, information on these changes during heat treatment allows for a deeper understanding of the patterns of physicochemical changes in the liquorice root extract that occur during evaporation, and provides a scientific basis for rational evaporation process conditions that lead to the production of a high-quality product.

The study of the thermal-physical properties of colloidal capillary-porous and dispersion media is conducted in two directions:

> the first direction involves the experimental determination of TPC using established methods based on thc solution of boundary value problems in heat conduction;

> Secondly, the analytical determination of thermal conductivity based on theoretical concepts regarding the mechanism of heat transfer in model structures characteristic of real solids and dispersed systems. These models cannot accurately reflect the structure of real bodies and therefore calculations using theoretical formulas are, to a certain extent, approximate, particularly as the TPC of liquorice root extract at various concentrations has not yet been determined,

The literature contains information on a whole range of methods and instruments for studying the thermal properties of various materials. Methods of experimental research into thermal properties can be classified according to a number of criteria:

1) the commonality of the underlying theory;
2) similarity of operating modes;

3) the amount of information obtained from a single experiment;
4) the temperature range of the tests;
5) the nature of the materials under investigation;
6) the principles of measuring temperature and heat quantity, **etc.**

Methods for measuring thermal properties can be divided into four main groups, using the commonality of theory and measurement principles as the classification criterion:

1) balance methods;
2) steady-state;
3) steady-state;
4) transient (non-steady-state) mode.

Each of these generally has limited application due to the specific thermal measurement properties of materials that differ in nature.

A description of the methods and apparatus for measuring the density of a substance is given in the work / /. Devices designed to measure density are called densitometers. Densitometers are classified according to their design and principle of operation as float, piezometric, weighing, diaphragm and radioisotopic types.

When selecting the method for determining TPCs, the following prerequisites were taken into account: ensuring the initial and boundary conditions of heat transfer in the experiment, corresponding to the specific technological process. The method used must be comprehensive, allowing three thermophysical characteristics to be determined in a single experiment, with a single sample and on a single instrument.

The experimental setup for determining TPC comprises a device for cooling and automatically inserting the probe into the material under investigation to a specified depth, an automatic self-recording single-point potentiometer KSP-4 with a measurement range of 0–100 °C, and an UT-15 ultra-thermostat. The rigidity of the structure prevents accidental movement of the sensor within the test material. This fulfils one of the key requirements of this TPC measurement method: precise knowledge of the sensor's location and the time of insertion. This device allows the measuring equipment to be switched on automatically the moment the sensor is fixed in the extract. It significantly facilitates experimental studies of the thermal physical properties of materials, has low inertia and provides good reproducibility of experimental results.

A schematic diagram of the device is shown in Fig. 2.

It consists of a base 1, a bracket 2, to which a frame 4 with a platform 5 is secured by a screw 3 (the latter can be adjusted in height depending on the specified depth of sensor fixation). Platform 5 holds a thermally insulated vessel 15 containing melting ice 14. A tube II, housing the sensing element 12, is welded into the centre of the vessel. For the convenience of conducting experiments, the thermally insulated vessel is made removable. It is secured to the plate by means of a support bracket 7, fixing screws 10 and a lever 8 with a spring 9. In the upper initial position, the temperature sensor 12, secured in the carriage 25, is held in place by the petals 21. When a pulse is applied to the electromagnet 19, the electromagnet's rod 20 rises, the petals 21 release the carriage 25, which, under the action of the weight 18, slides along the guides 17 mounted on the frame 23. The carriage 25 moves the sensor into the test material 24. As it moves, the temperature sensor pierces the sheet of paper 16, which is provided to prevent convective heat flows in the tube 11. In the lowest position, the platform with the load 18 activates a microswitch 22, which activates a secondary measuring instrument, for which a KSP-4 electronic single-point self-recording potentiometer was used. The cold junctions of the thermocouples are thermostated; they are placed in a test tube 13 situated in water with melting ice. The paper / / provides a detailed description of the method and apparatus. The samples were heated in a double-walled cylindrical vessel with a thermal insulation layer in a water bath. Measurements were taken after the sample reached a specified temperature during the heating process. The time interval between two measurements depended on the cooling time of the temperature sensor in the thermally insulated vessel in the ice-water bath.

The paper / / sets out the theoretical foundations of this method, in which a solution is given to the boundary value problem of heat conduction in a spherical solid body under fourth-order boundary conditions. The aim of the problem was to determine the body's temperature as a function of time given its known thermal inertia properties.

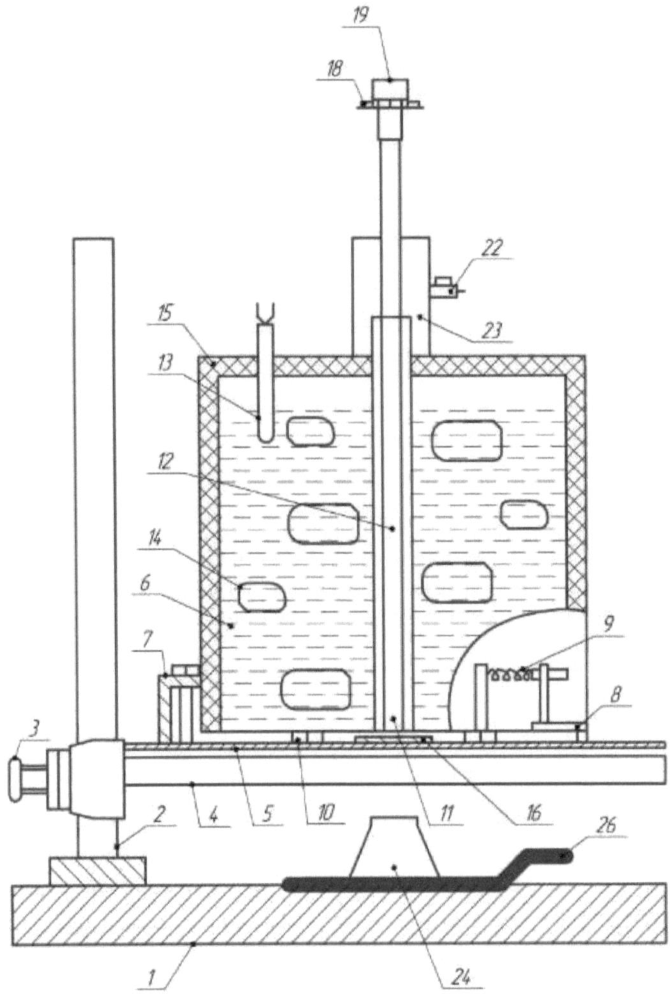

***Fig. 2** Schematic diagram of the setup for determining the thermophysical characteristics of liquorice root extract*

In the paper / /, a method for calculating the thermal-physical characteristics of a product based on this experimental method was developed.

The thermal-physical characteristics are investigated as follows.

First, the sensor is calibrated and its characteristics determined. The sensor is then inserted into the material under investigation, and the sensor's heating curve within the product is obtained as follows:

$$\theta = f(\tau) \qquad \textbf{(2.1)}$$

where: Θ *is* the dimensionless temperature.

$$\theta = \frac{T_1 - T_2}{T_c - T_0}$$

Here: T_i and T_s , T_2 **and** T_o - respectively, the current and initial temperatures of the medium under investigation and the probe; the thermal conductivity coefficient
is determined by the equation

$$\lambda = \frac{\beta}{S} \qquad \textbf{(2.2)}$$

where: S *is* the area bounded by the curve, m^2;
β *is* the sensor constant, calculated using the equation

$$\beta = \frac{r^2}{m} \qquad \textbf{(2.3)}$$

where: r – probe radius, m;
t *is* the thermosensor constant characterising its heat storage capacity, calculated using the formula

$$m = \left(\frac{c_2 \cdot \rho_2}{3}\right)^{-1} \qquad \textbf{(2.4)}$$

where: $c_2 \cdot \rho_2$ - is the volumetric heat capacity of the probe, $\frac{кДж}{м^3 \cdot K}$
To determine the remaining thermal characteristics, the experimental curve $\theta = f(\tau)$ is processed and presented as a functional relationship

$$\theta_2 = \sqrt{\tau^3} = f\left(\frac{1}{\tau}\right) \qquad \textbf{(2.5)}$$

where: θ_2 is a dimensionless complex quantity calculated using the equation

$$\theta_2 = \frac{T_i}{T_{max}} \qquad \textbf{(2.6)}$$

where: T_g - temperature of the current measurement on curve (I), K;
T_{max} - maximum probe heating temperature, K;
τ - time interval between measurements, s.
From curve (2.5), the value is determined; then, using equation (2.8), , the value of the criterion K is determined, which defines the ratio of the specific heat capacities of the material under investigation and the probe

$$K = m \cdot c_1 \cdot \rho_1 \qquad (2.7)$$

Using the converse theorem followed by expanding the resulting function into a series, we obtain

$$\theta_2(\tau) = \frac{1}{2 \cdot K \cdot \sqrt{\pi} \cdot \sqrt{F_0^3}} \cdot \left\{1 + \frac{3 \cdot (2-K)}{2 \cdot K \cdot F_0} + \frac{15 \cdot \left[(K-2)^2 - 1\right]}{4 \cdot K^2 \cdot F^2} + \frac{105 \cdot (K-2) \cdot \left[(K-2)^2 - 2\right]}{8 \cdot K^3 \cdot F_0^3} + \ldots\right\} \qquad (2.8)$$

Substituting the found value into equation (2.9), the Fourier criterion is calculated

$$\frac{1}{F_{0\max}} = -\frac{(2-K) \cdot K}{S \cdot \left[(K-2)^2 - 1\right]} \qquad (2.9)$$

Taking into account the found values of the criterion and criterion K at the moment of reaching the maximum temperature, the TFC complex of the licorice root extract under study is determined using equations (2.10, 2.11, 2.12): thermal conductivity coefficient, *m²/s*

$$\alpha_1 = \frac{F_{0\max} \cdot r^2}{\tau_{\max}} \qquad (2.10)$$

specific heat capacity, *kJ/kg·K*

$$c_1 \cdot \rho_1 = \frac{K}{m} \qquad (2.11)$$

thermal conductivity *coefficient, W/m²·K*

$$\lambda = \alpha_1 \cdot c_1 \cdot \rho_1 = \frac{\beta}{S} \qquad (2.12)$$

In general, the error of the method is due to the error in determining the area *S of* the bounded curve *Θ* = *f*(t), which depends on the accuracy class of the recording instrument and the error in measuring the ordinate *Ati* on the chart.

To determine the density of the liquorice root extract, we used:

- VLK-500 laboratory analytical balances;
- 100 ml conical flasks;
- a thermometer for measuring temperature;
- thermostats for heating the extract to a specific temperature.

The experiments were carried out in the following order: a volume *V* of extract of a specific concentration was poured into a 100 ml conical flask.

The flask containing the solution was weighed on the VLK-500 analytical balance and heated in a thermostat to the required temperature / /. After heating to the required temperature, the flask containing the heated solution was weighed. Knowing the mass and volume of the heated solution, the density of the extract was determined using the formula:

$$\rho_{eks} = \frac{m}{V} \quad \frac{кг}{м^3} \qquad \textbf{(2.13)}$$

The experiments were carried out within the investigated range of temperatures (25–75 °C) and concentrations (15–65 %).

2.2. Experimental determination of the TFC of liquorice root extract root

As noted above, the TPPs of products have a significant influence on the heat treatment process. For scientifically sound calculations of new technological processes and apparatus and equipment in the food industry, it is necessary to have reliable experimental data on the thermophysical properties of food products as a function of their temperature and concentration.

In view of the growing demands for improved extract quality, there is an increasing need to investigate its thermophysical properties. Without knowledge of these properties, it is impossible to conduct analytical research or develop methods for the engineering calculation of the evaporation process. The availability of data on thermophysical properties allows for the justification of the selection of optimal operating parameters for the evaporation process and, consequently, the creation of evaporators that are most energy-efficient.

TPH provide a quantitative assessment of the product's thermophysical properties. They determine the heat storage capacity of products and the rate at which they heat up or cool down.

The TPPs of the extract are most strongly influenced by concentration and temperature; therefore, experiments were conducted within a temperature range of 25°C to 75°C and a concentration range of **15% w/w** to **65%** w/w.

Solutions of varying concentrations were prepared as follows: distilled water and concentrated extract were mixed in specific ratios, and the dry matter content was determined using an RPL-2 refractometer. To determine the dry matter content in the extract using the refractometric method, the dependence of the extract's refractive index on the content of extractable substances was experimentally established.

To bring the experimental conditions closer to production conditions and prevent the possibility of increased data scatter due to changes in the mass dispersion, the prepared sample was heated once in a UT-15 thermostat to the required temperature, and a series of measurements was carried out. During the measurements, a sensor at a temperature of^0°C was instantly immersed in the sample, and its heating was automatically recorded by a KSP-4 potentiometer. Once a steady-state temperature was reached, the sensor was removed from the sample.

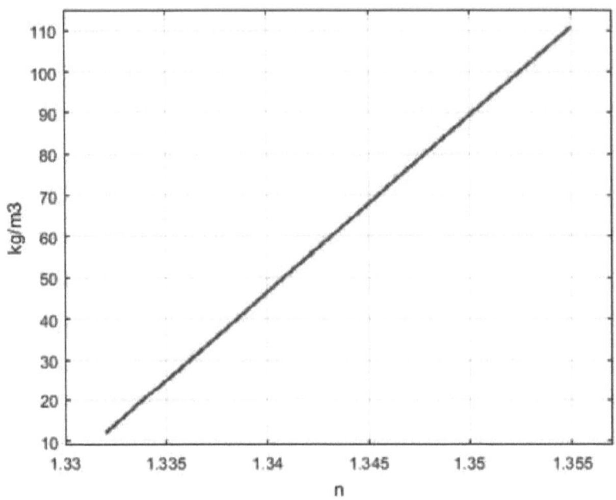

Fig. 2.1. *Dependence of the refractive index of liquorice root extract on extractable substances.*

To obtain an empirical equation of the form

$$y = (t, CB) \qquad \textbf{(2.13)}$$

where: y – the coefficients /2/:

λ is the thermal conductivity, W/m·K;

a – thermal diffusivity, ^{m3}/s;

z

avg is the specific heat capacity, kJ/m K;

ρ – density of liquorice root extract, $^{kg}/m^3$;

SB – concentration of liquorice root extract

t – process temperature, °C

A six-level factorial design was carried out with a temperature range from^{25}°C to 7^{5}°C and a concentration range from 15% to 65%. Based on the analysis of experimental data using the least squares method, an equation in the form of second-degree polynomials was obtained:

$$y = b_0 + b_1 \cdot CB + b_2 \cdot t + b_3 \cdot CB \cdot t + b_4 \cdot CB + b_5 \cdot t^2 \qquad \textbf{(2.14)}$$

The programmes developed to determine the corresponding coefficients of the equation allowed for the calculation of the correlation coefficient *S*. The estimates of the coefficients of the equation *y=(t,CB)* for the coefficients of thermal conductivity, temperature conductivity, specific heat capacity and density are given below.

The equation and correlation coefficients for determining the thermal conductivity coefficient λ, *W/m K,* are:

ά 0 = 0.715

Ђ. = 0.316 -10 "

Ђ 2 = 0.645 · 10 -ᴵ

b_3 = 0.263 · 10 '

b_4 = 0.135 × 10 '

b_5 = 0.362 × 10 '

S = 0.9689929

The equation and correlation coefficients for determining the thermal conductivity coefficient 10^6 , m^2 /s are:

ό 0 = 0.710

B_x = 0.147 × 10⁻

B_2 = 0.183 × 10⁻ᴵ

b_3 = 0.661 × 10

b_4 = 0.523 × 10 '

b_5 = 0.391 × 10 "

S = 0.9811622

The coefficients of the equation and correlation for determining the specific *heat* capacity *of* the liquorice root extract are:

ό0 = 4232

b_x = -0.338 -10 '

b_2 = -0.1283 -10 "

b_3 = -0.6839 -10

b_4 = -0.3827 -10 '

b_5 = -0.1504

5 = 0.9810658

The equation and correlation coefficients for determining the density; 10^3, $^{kg}/m^{3}$ of liquorice root extract are:

b_0 = 1.0983

$Ъ_x$ =-0.6826 -10 "

$Ъ_2$ = -0.7187 -10

b_3 = -0.3545 -10

b_4 = -0.1733 -10 '

$Ъ_5$ = -0.3553 -10 '

5 = 0.9722788

The numerical values of the coefficients were calculated using the MatLab software package. Based on the averaged data (Tables 1.1–1.5), graphs showing the dependence of the TFH of liquorice root extract on temperature and concentration were plotted; these are presented in Figures 2.2–2.6.

Table 1

Experimental values of the thermal conductivity coefficient of liquorice root extract $(10^{-3} \cdot Bm/(м^2 \cdot K))$

Temperature, °C	Concentration (w/v), %					
	15	25	35	45	55	65
25	678	659	639	615	582	534
35	693	665	644	626	596	539
45	719	673	658	639	609	545
55	727	680	670	653	621	553
65	735	688	679	668	637	569
75	748	699	692	680	652	576

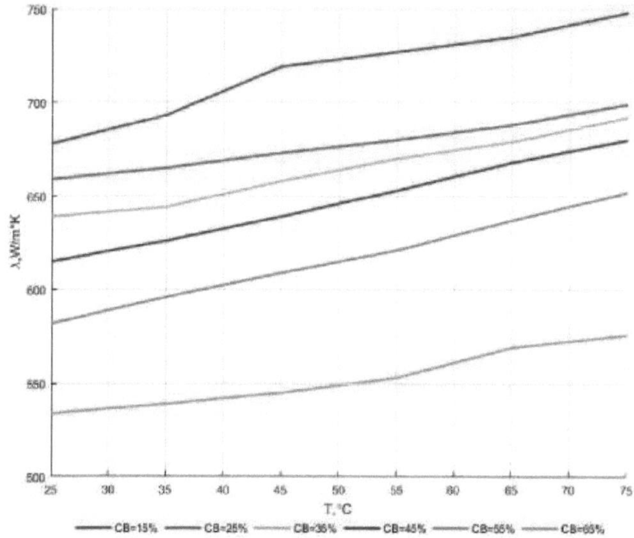

Fig. 2.2. ***Dependence of the thermal conductivity coefficient of liquorice root extract on temperature at various concentrations of the evaporated extract***

Table 2

Experimental values of the thermal conductivity of liquorice root extract (10^{-6} *W·m·s)*

Temperature, °C	Concentration (w/v), %					
	15	25	35	45	55	65
25	0.165	0.163	0.160	0.159	0.157	0.147
35	0.169	0.167	0.165	0.163	0.160	0.148
45	0.175	0.172	0.172	0.169	0.166	0.155
55	0.179	0.175	0.175	0.172	0.170	0.157
65	0.182	0.181	0.182	0.180	0.171	0.159
75	0.188	0.185	0.185	0.184	0.176	0.162

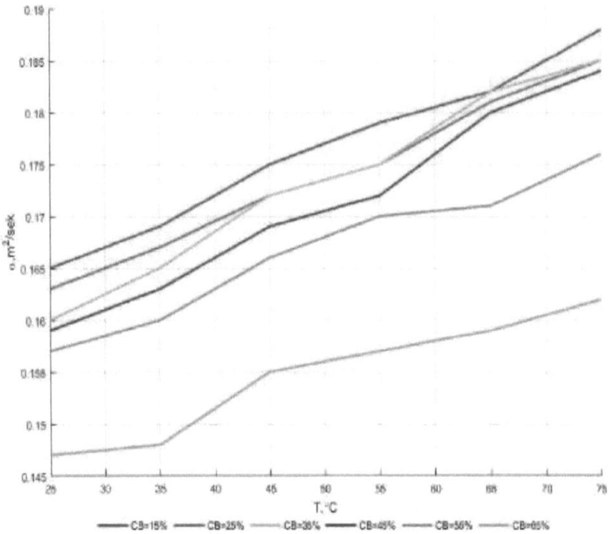

Fig. 2.3. ***Dependence of the temperature-conductivity coefficient of liquorice root extract on temperature at various concentrations of the evaporated extract***

Table 3

Experimental values of the volume heat capacity of liquorice root extract (kJ/(m³·K))

Temperature, °C	Concentration (w/v), %					
	15	25	35	45	55	65
25	4,092.7	4121.7	4026.4	3,830.9	3566.9	3842
35	4092.2	4076.5	4012.8	3780.5	3547.7	3,373.1
45	4091.7	4,034.6	3,943.9	3718.3	3,535.2	3,267.2
55	4,057.2	3967.7	3940.6	3,714.9	3529.5	3,254
65	4023.4	3,950.7	3938.3	3667.4	3518.7	3,246.2
75	3963.3	3924.3	3927.5	3660.1	3518.6	3240.2

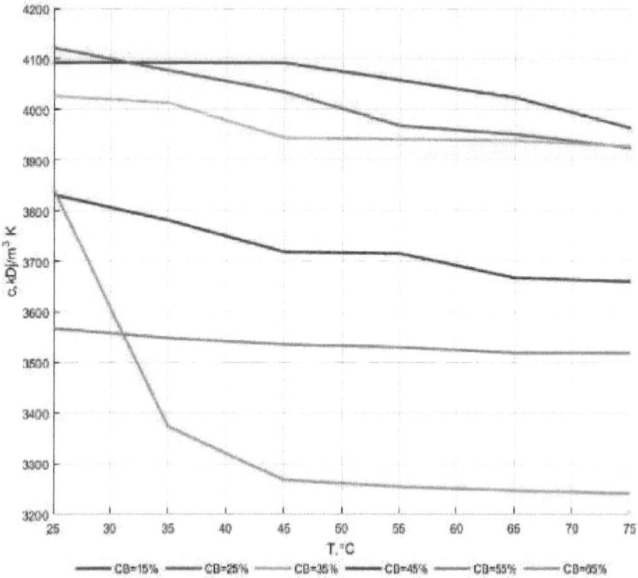

Fig. 2.4. ***Dependence of the specific heat capacity of liquorice root extract on temperature at various concentrations of the evaporated extract***

Table 4

Experimental values of the specific heat capacity of liquorice root extract $^{(J/kg\cdot°C)}$

	Concentration (SV), %					
Temperature, °C	15	25	35	45	55	65
25	3,812	3581	3361	3097	2849	2601
35	3825	3598	3375	3109	2875	2625
45	3842	3612	3497	3122	2893	2637
55	3859	3631	3509	3135	2905	2652
65	3865	3645	3522	3108	2925	2674
75	3878	3665	3546	3123	2942	2689

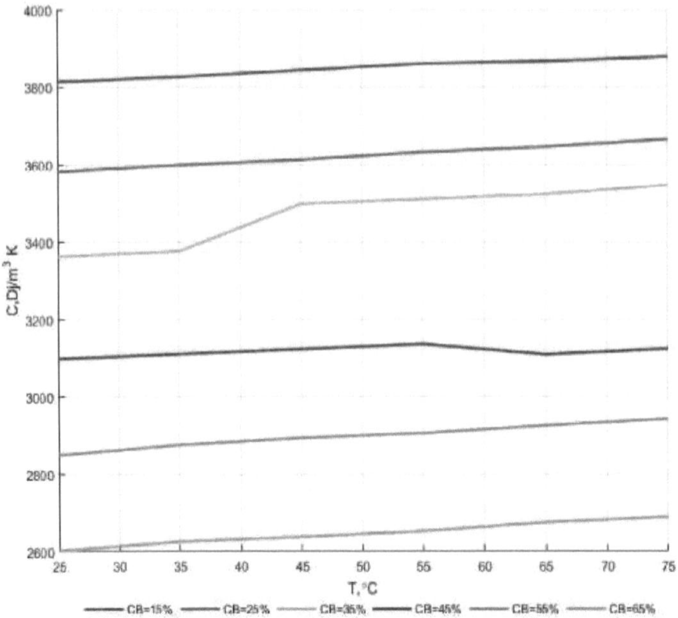

Fig. 2.5. ***Dependence of the specific heat capacity of liquorice root extract on temperature at various concentrations of the evaporated extract***

Table 5

Experimental values of the density coefficient of the LICORICE ROOT (kg/m³)

Temperature, °C	Concentration (w/v), %					
	15	25	35	45	55	65
25	1073	1151	1198	1237	1252	1285
35	1070	1133	1189	1216	1234	1257
45	1065	1117	1185	1191	1222	1239
55	1053	1090	1180	1185	1215	1227
65	1041	1079	1175	1180	1203	1214
75	1022	1058	1164	1172	1196	1205

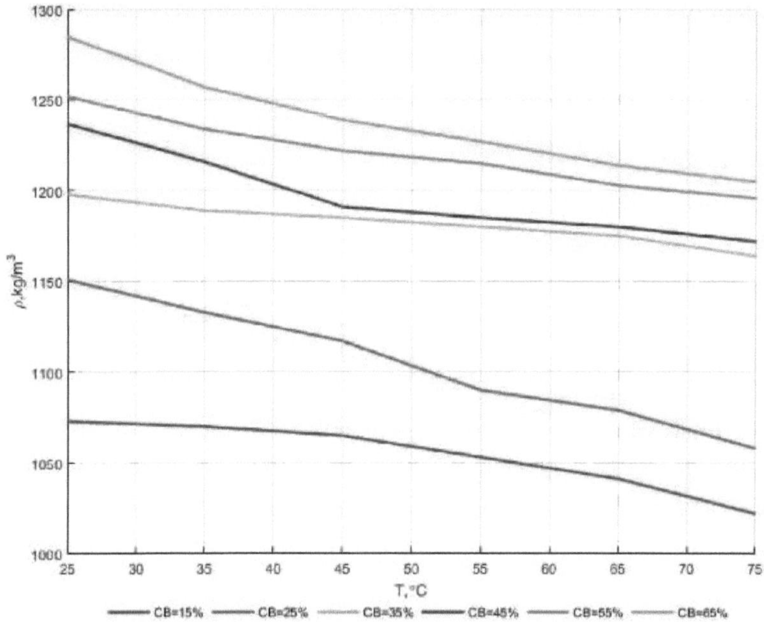

Fig. 2.6. *Dependence of the density of liquorice root extract on temperature at various concentrations of the evaporated extract*

Analysis of how thermophysical properties change with temperature at various concentrations shows that the thermal conductivity and temperature conductivity of the liquorice root extract decrease with increasing concentration and increase with increasing temperature. The viscosity *of* the extract decreases with increasing concentration and temperature; the density of the liquorice root extract decreases with increasing temperature and increases with increasing concentration.

The studies carried out have identified the main patterns of change in the TFC of liquorice root extract during heat treatment and have proposed empirical relationships between the TFC of the extract and temperature (25°C 75°C) and concentration (15% 65%), which can be used for calculation and in the analysis of the evaporation process.

2.3. Study of the rheological properties of solutions

Food products undergo thermal and mechanical processing during technological processes. As a result , the structure of food dispersive systems is partially or completely destroyed and their rheological properties change. Structural-mechanical characteristics are used to calculate the flow of products through machine working parts in order to determine their mechanical parameters (geometric, kinematic and dynamic); they reflect essential aspects of product quality.

Furthermore, structural-mechanical characteristics are taken into account when calculating various physical processes: hydromechanical, thermal (conductive, high-frequency and other types of heating, evaporation), mass transfer, etc. Along with other food products, liquorice root extract possesses anomalous viscosity, the value of which depends significantly on the flow velocity gradient, temperature and concentration. Therefore, viscosity data are of great importance in the development of processing equipment / /.

The viscosity of the liquorice root extract was measured using a 'Reotest-2' rotational viscometer. In terms of its measuring principle, the 'Reotest-2' is a structural-rotational viscometer. Its application extends to a wide range of non-Newtonian fluids. The advantages of the instrument include the ability to adjust the rotor speed over a wide range and the automatic recording of torque readings, which simplifies the calculation of rheological parameters, the presence of an electromechanical drive and a set of measuring rotors and cups, and a constant product height in the cups.

The viscometer rotor speed is selected using a switch drive. Twelve different speeds can be selected with a step factor of ' ,

The viscosity of the extract depends to a large extent on temperature and concentration. Therefore, viscosity measurements were carried out in parallel with temperature determination. This factor has been taken into account in this viscometer. A liquorice root solution with varying dry matter content was prepared for the experiments using the method described in section 2.1.2.

In the experiments, a viscometer measuring device (- , fig.) was used, which allows the effective viscosity to be determined in the range (0.03–1.2) × 10 *Pa·s,* with a radius ratio of 0.94 (Table I). It was used to measure viscosity in the flow rate gradient range from 1 to 437 $^{s^{-1}}$,in the temperature range from 25 °Cto 75 °C and at concentrations ranging from 15% to 65% SV. The instrument's measurement error is ±3%.

The shear stress Q was calculated using the formula:

$$Q = z \cdot \alpha \qquad \textbf{(2.14)}$$

where: Z *is* the constant of the rotating cylinder;

a – the instrument's reading on the scale.

Table 2

Viscosity meter measuring device	Cylinder radius ratio	Range	Shear stress, Pa	Deformation rate, E, s^{-1}	Effective (dynamic) viscosity	
					General limits	Preferably
	0.92	**1**	**12–120**	**0.5–437**	**0.03–240**	**0.1–200**
		2	**60–600**		**0.15–1200**	

Based on the obtained values of shear stress and strain rate gradient, flow curves of the liquorice root extract were plotted as a function of temperature and concentration; the effective viscosity $\eta_{эф} = \frac{Q}{\gamma}$ and a model describing the flow behaviour of the liquid were determined, for which the consistency coefficient of the extracts and the flow index n were calculated.

2.4. Experimental determination of the rheological properties of liquorice root extract

A study of the rheological properties of liquorice root extract showed that the extract belongs to the class of pseudoplastic fluids, which can be described by the Ostwald–de Vill equation:

$$Q = K \cdot \gamma^{n} = \eta_{эф} \cdot \gamma \qquad \textbf{(2.15)}$$

where:

Q - shear stress, Pa;

K – a constant characterising the consistency of the product;

γ – relative strain under shear;

γ – shear strain rate; η – shear modulus.

Studies using a rotational viscometer yielded flow curves $Q = f(\gamma, t)$ and viscosity-rate characteristics $\eta_{3\varphi} = f(\gamma/t'^{\,S})$ for liquorice root extract over a temperature range $^{of\ 25°C}$to 75°C ^{0} C and a concentration range of 15%–65%.

Fig. 2.8 - 2.13 show the viscosity-shear rate characteristics of the extract. At high temperatures, when the product structure is more severely disrupted, the increasing flow rate gradient of the medium has less effect on viscosity, and *the* $Q = f(\text{u}')$ curves lie lower as the temperature rises. The viscosity-shear rate characteristics of the extract, plotted on a logarithmic scale, proved to be linear, and the consistency constant and shear index were determined for each temperature value using the equation lg $\eta_{3\varphi}$ $= \lg(\text{M} \cdot \gamma^{m}$ *)*. Based on these data, the relationships between *K* and η as a function of the temperature of the liquorice root extract were plotted.

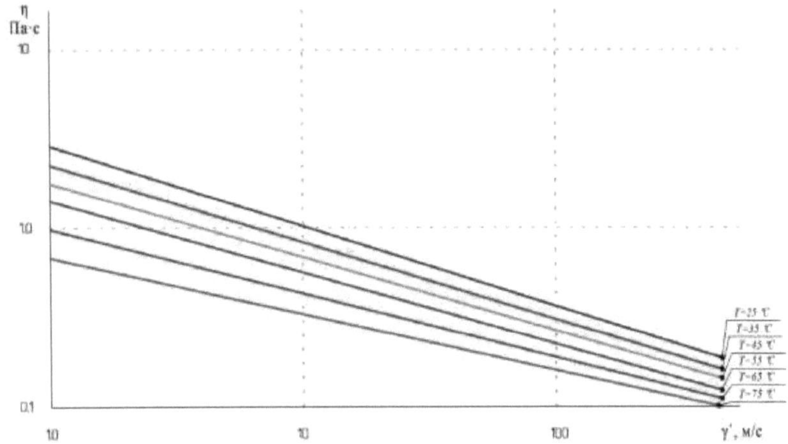

***Fig.** 2.7. Dependence of the effective viscosity of liquorice root extract on the shear rate gradient and heating temperature at 5% SV.*

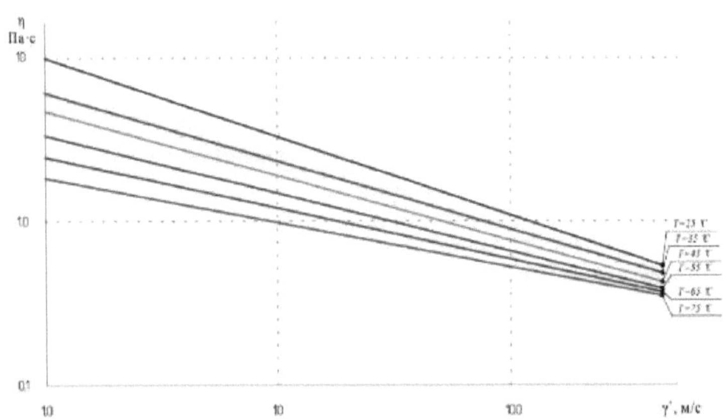

***Fig. 2.8.** Dependence of the effective viscosity of the liquorice root extract on the shear rate gradient and heating temperature at SV = 5%.*

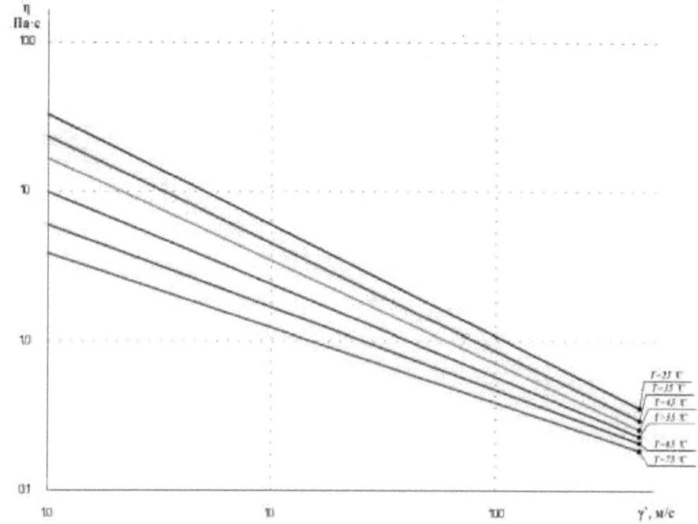

Fig. 2.9. *Dependence of the effective viscosity of liquorice root extract on the shear rate gradient and heating temperature at SV = 35%.*

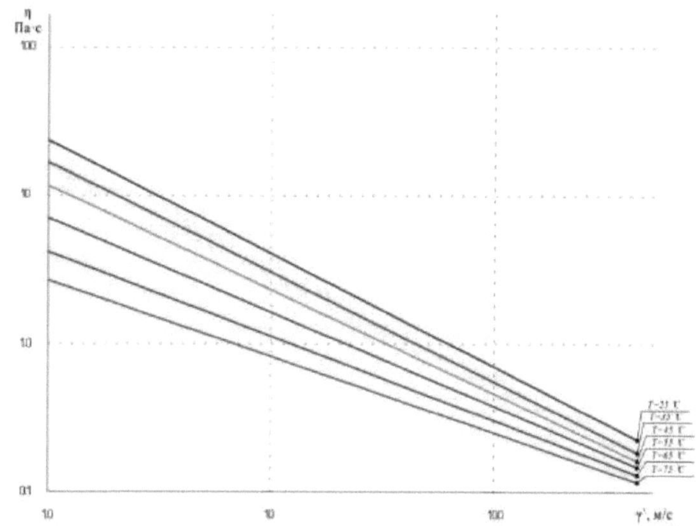

Fig. 2.10. *Dependence of the effective viscosity of liquorice root extract on the shear rate gradient and heating temperature at SV = 45%.*

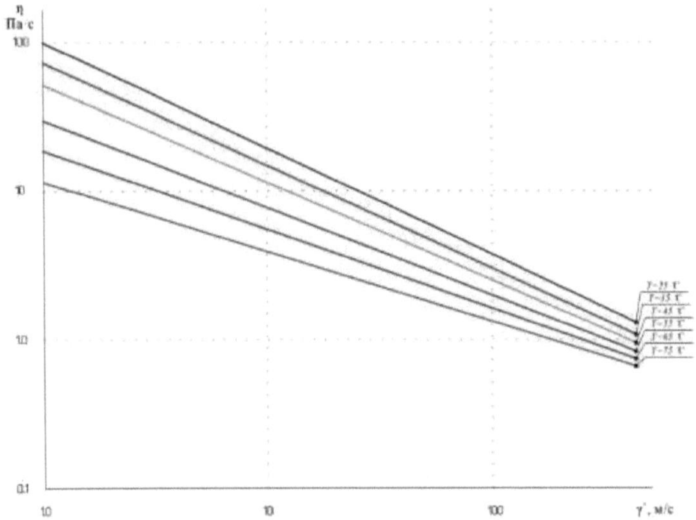

Fig. 2.11. *Dependence of the effective viscosity of liquorice root extract on the shear rate gradient and heating temperature at SV = 55%.*

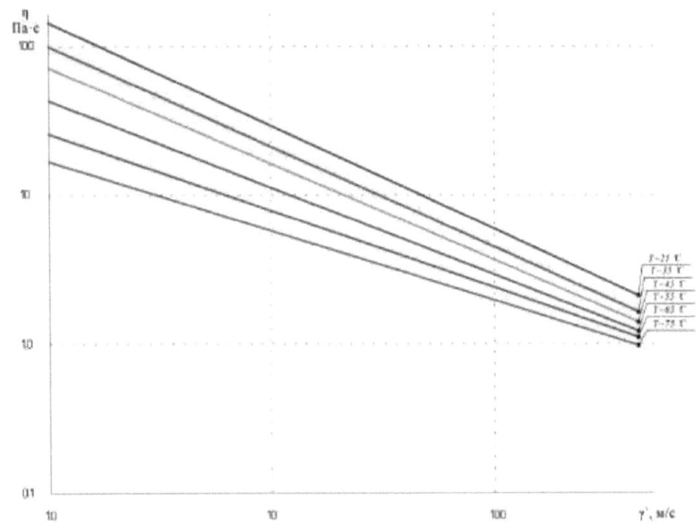

Fig. 2.12. *Dependence of the effective viscosity of liquorice root extract on the shear rate gradient and heating temperature at DS = 65%.*

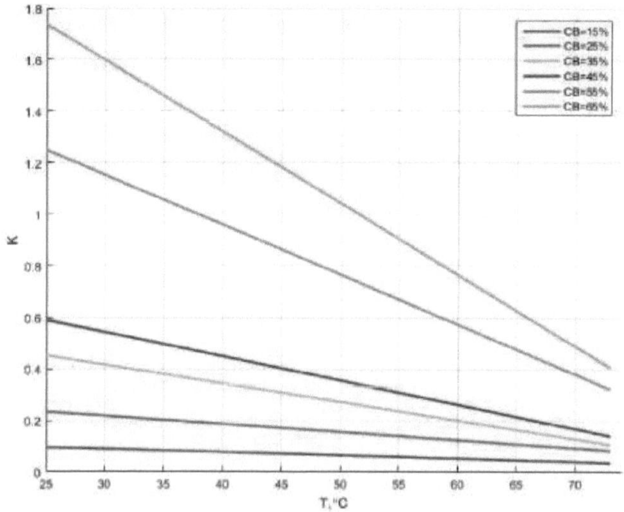

Fig. 2.13. Dependence of the consistency coefficient K of liquorice root extract on temperature at various dry matter contents of the evaporated extract

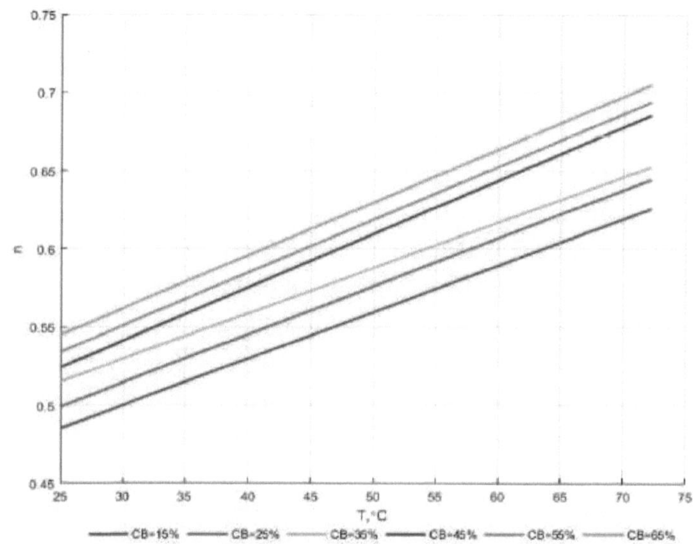

Fig. 2.14. Dependence of the flow index of liquorice root extract on temperature at various concentrations of the evaporated solution

Analysis of the flow curves and viscosity-velocity characteristics shows that as the temperature rises and the flow velocity gradient increases, the viscosity of the media decreases, whilst it increases with increasing concentration. This pattern of change in shear stress and effective viscosity with respect to the velocity gradient allows us to conclude that the extract belongs to the class of non-Newtonian pseudoplastic fluids.

The values of the parameters *K and η_p from* the regression equations $K = f(t)$ and $\eta = /(i)$, obtained by the least squares method, are given in Table 3. As can be seen from the results presented, the consistency constant K of the extract decreases with increasing temperature but increases with increasing concentration. The higher the dry matter content in the extract, the more viscous it becomes. The viscosity index η is less than one, which means that the extract belongs to the class of pseudoplastic liquids. The viscosity index η increases with rising temperature and decreases with increasing concentration.

Thus, rheological studies have established that liquorice root extract is a pseudoplastic fluid; the values of the product's consistency constant and flow index ( ) have been determined, allowing conclusions to be drawn regarding changes in its structure during processing. The viscosity-shear rate relationships obtained can be used in the design of process equipment for the production of liquorice root extract.

Table 3.

Values of the Ostwald–de Villiers equation coefficients for liquorice root extract

Concentration, %	Temperature, °C	Consistency coefficient	Flow index
15	**25**	**0.0954**	**0.485**
15	**35**	**0.0824**	**0.502**
15	**45**	**0.0727**	**0.518**
15	**55**	**0.0599**	**0.557**
15	**65**	**0.0500**	**0.586**
15	**75**	**0.0352**	**0.627**
$K = 0{,}1208 - 0{,}001 \cdot t$ $\quad n = \frac{t}{-0{,}8845 + 1{,}4704 \cdot t - 0{,}0016 \cdot t^2}$ $\quad \sigma = 1{,}75\%$ $\quad \sigma = 0{,}65\%$			
25	**25**	**0.233**	**0.499**
25	**35**	**0.170**	**0.517**
25	**45**	**0.139**	**0.530**
25	**55**	**0.116**	**0.568**

25	**65**	**0.091**	**0.602**
25	**75**	**0.076**	**0.645**
$K = 0{,}005 + \frac{5{,}7812}{t}$ $n = \frac{t}{1{,}3891 + 1{,}4487 \cdot t - 0{,}0033 \cdot t^2}$ $\sigma = 4{,}5\%$ $\sigma = 0{,}74\%$			
35	**25**	**0.451**	**0.515**
35	**35**	**0.349**	**0.529**
35	**45**	**0.275**	**0.549**
35	**55**	**0.193**	**0.574**
35	**65**	**0.133**	**0.638**
35	**75**	**0.094**	**0.654**
$K = 0{,}7522 - 0{,}0136 \cdot t$ $n = 0{,}4697 + 0{,}0011 \cdot t$ $\sigma = 2{,}13\%$ $\sigma = 1{,}432\%$			
45	**25**	**0.589**	**0.524**
45	**35**	**0.482**	**0.530**
45	**45**	**0.359**	**0.563**
45	**55**	**0.247**	**0.608**
45	**65**	**0.171**	**0.643**
45	**75**	**0.126**	**0.686**
$K = 3{,}1722 - 0{,}0664 \cdot t + 0{,}0003 \cdot t^2$ $n = t \cdot \left(-130{,}85 + 55{,}366 \cdot \frac{\ln}{t} \right)$ $\sigma = 1{,}144\%$ $\sigma = 0{,}64\%$			
55	**25**	**1.25**	**0.534**
55	**35**	**1.004**	**0.552**
55	**45**	**0.824**	**0.568**
55	**55**	**0.555**	**0.615**
55	**65**	**0.381**	**0.653**
55	**75**	**0.274**	**0.694**
$K = \frac{0{,}1866 \cdot t}{\exp(0{,}527)}$ $\eta = 0{,}5466 - 0{,}017$ $\sigma = 2{,}92\%$ $\sigma = 1{,}104\%$			
65	**25**	**1.738**	**0.545**
65	**35**	**1,291**	**0.557**
65	**45**	**1.046**	**0.597**
65	**55**	**0.571**	**0.652**
65	**65**	**0.424**	**0.672**
65	**75**	**0.316**	**0.705**

$$K = 3{,}1722 - 0{,}0664 \cdot t + 0{,}0003 \cdot t^2 \qquad n = \frac{0{,}4575 \cdot t}{\exp(0{,}0058 \cdot t)}$$

$$\sigma = 2{,}92\% \quad \sigma = 1{,}104\%$$

2.5. Experimental determination of boiling point

2.5.1. Investigation of the boiling point of solutions

The boiling point is the main indicator of extract concentration, characterising the state of equilibrium of systems at specific pressures within the apparatus; at constant pressure, the boiling point of a solution depends on its concentration.

The boiling point of solutions can be determined using Raoult's law. The boiling point of plant solutions and the vapour pressure of the solvent above it deviate from the patterns characteristic of ideal solutions. Therefore, various empirical formulas must be used for calculations. If the boiling point of a pure solvent $_{t(p)}$is known, then the boiling point of plant material can be expressed as:

$$t_k = t_p + \Delta t \qquad \textbf{(2.4.1)}$$

where: $\Delta\zeta$ - temperature depression.

The boiling point of solutions is an important parameter in the design and calculation of evaporators and depends on the chemical nature of the solid and the solvent; it increases with increasing concentration and external pressure on the liquid.

If the boiling point of a solution of a given concentration is known at only one pressure, the boiling point of this solution at any given pressure in the apparatus can be determined. The calculation is performed using Babo's rule: where: P_{pl} , P_p 2 – the saturated vapour pressure of the solution at different temperatures;

$$\left(\frac{P_{p1}}{P_1}\right)_t = \left(\frac{P_{p2}}{P_2}\right)_t = const \qquad \textbf{(2.4.2)}$$

Pi, P_2 - the saturated vapour pressure of the pure solvent at the same temperatures.

Babau's law applies to dilute solutions, whilst for concentrated solutions, V.N. Stabnikov established corrections / /.

The boiling point of a solution at various pressures can be determined using the rule of linearity of chemical-technical functions: the ratio of the difference in boiling points of a given liquid (^{zz} $_{-}^{tz}$) at two arbitrarily chosen pressures to the difference in boiling points of another liquid $^{v\ e}$ at the same pressures is a constant. Water is the second liquid for which boiling points at various pressures are known.

2.5.2. Description of the experimental setup

A diagram of the experimental setup for determining the boiling point is shown in Fig. 2.14. The setup consists of a boiling tube 1 made of glass. The diagram includes vessel 2 for the extract, compressor 3 for feeding the extract from vessel 2 into boiling tube 1, a regulator 4 for maintaining a constant flow rate of the extract, pressure gauge PE (*item 1-a*) for monitoring the constant pressure in vessel 2, a valve 5 for regulating the flow rate of the extract, a thermometer TE *(item 3-a*) for determining the boiling point of the liquorice root extract, a vacuum pump 6 for determining the vacuum in the boiling tube 1, valve 7 – for regulating the vacuum, trap 8 – for condensing the vapour, vessel 9 – for determining the volume of collected condensate, monovacuum gauge PE, *item 2-a* – for determining the vacuum, separator 10, made of glass. Vessel 9 is used to determine the volume of separated liquid.

The experimental setup operates as follows: an extract of a specific concentration is continuously fed into the boiling tube. In this process, the compressor 3, the reducer 4, the vessel 2 and the valve 5 ensure a continuous and uniform supply of the extract. At the same time, the boiling apparatus 12 and the vacuum pump 6 are switched on. A specific vacuum is established using valve 7. Once the vacuum is established, the liquid begins to boil. The boiling point of the extract is determined using thermometer TE, *item 3a*.

2.5.3. Experimental procedure

Experiments to determine the boiling point of liquorice root extract as a function of pressure (vacuum) in the apparatus were carried out in a setup, the diagram of which is shown in Fig. 2.15. The boiling point of the liquorice root extract was determined at various concentrations and pressures in the apparatus.

The experiments were carried out in the following sequence: the boiling tube was filled with the extract of the appropriate concentration to 1/3 of its height. Then, using compressor 3 and pressure reducer 4, the required pressure value was set in vessel 2, ensuring the desired flow rate of the concentrated extract through the pipeline and valve 5 into the boiling tube. Slowly opening tap 5, the boiling tube is filled with the solution to 1/3 of its height. Using tap 5, the flow rate of the liquid is adjusted so that the level of the extract in the boiling tube remains constant. Once a steady-state flow of the extract is achieved, the boiler and the vacuum pump are switched on simultaneously.

In the boiling tube, the extract is heated to the boiling point determined by the vacuum set for this experiment using a monovacuum gauge; a specific vacuum is established using the vacuum pump, the maximum at which there is no significant carry-over of solution droplets of a given concentration.

Maintaining the solution level in the vessel, the experiment was carried out for 0.5–1 hour. The boiling point was measured using a thermometer, and the vacuum was measured using a monovacuum gauge. The experiments were repeated in the order described above for the following concentrations of liquorice root extract: 10, 15, 25, 30, 35, 40 and 40% w/v. Each experiment was carried out in triplicate under steady-state conditions.

2.5.4. Experimental results

The experiments were carried out at various pressures in the apparatus and at different concentrations of liquorice root extract in the range of 10–40% w/v. Table 4 presents the experimental data obtained on the experimental setup. The experiments were carried out under steady-state conditions in triplicate. Based on the experimental data, a plot was constructed showing the dependence of the boiling point of liquorice root extract of various concentrations on the pressure in the apparatus (Fig. 2.15). Analysis of the curves shows that the boiling point of the liquorice root extract increases with increasing pressure. The higher the concentration of the extract, the higher the boiling point at a given pressure in the apparatus.

To derive an empirical equation of the form, an experimental design was carried out over a concentration range of 10% to 40% w/v at various pressures.

$$t_k = f(CB, P) \qquad (2.4.3)$$

where: 4 – boiling point of the liquorice root extract, °C;

SV – concentration of the liquorice root extract, %;

P – pressure in the apparatus, Pa;

Based on the analysis of experimental data using the least squares method, the following equation was obtained:

$$T_k = b_0 + b_1 \cdot CB + b_2 \cdot P + b_4 \cdot CB \cdot P \qquad (2.4.4)$$

The developed programme provides the capability to calculate the correlation coefficient *S*. The resulting regression equation for determining the boiling point of the liquorice root extract is as follows:

$$T_k = \frac{31{,}027 + 1{,}76 \cdot P - 0{,}014 \cdot P^2}{69{,}45 \cdot \left(0{,}015 - 21{,}62 \cdot 10^{-6} \cdot CB - 9{,}2 \cdot 10^{-8} \cdot CB^2\right)} \qquad (2.4.5)$$

Equation /2.4.5/ can be used in thermal calculations of the evaporation process .

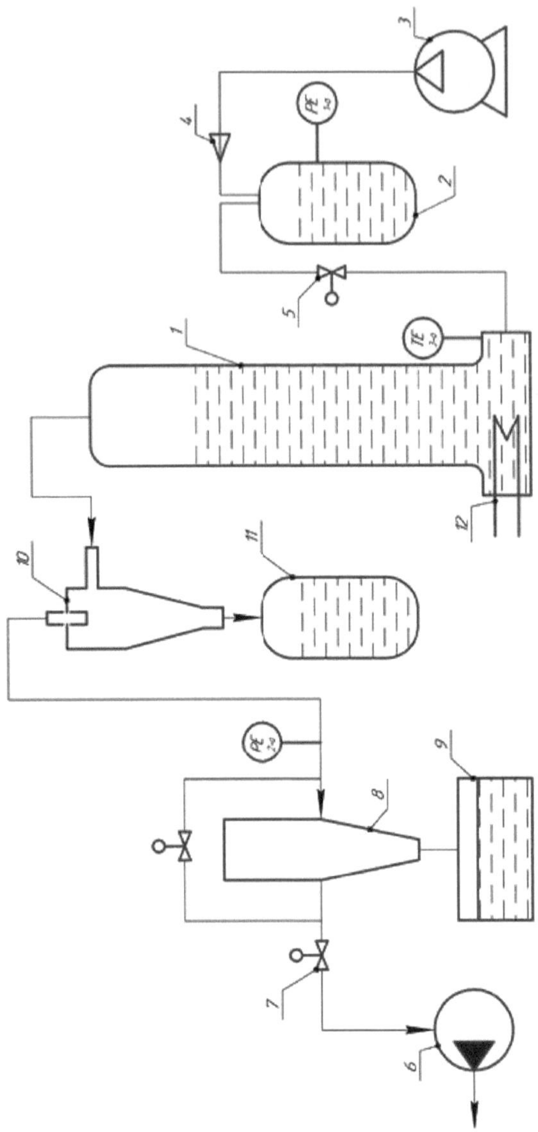

Fig. 2.15. **Schematic diagram of the experimental setup for determining the boiling point of the extract**

Table 4

Experimental values of the boiling point of liquorice root extract as a function of its concentration and vacuum

Vacuum, mm	Concentration of SV, %

Hg	10	15	25	30	35	40
350	82	82.5	83	83.5	84	85
500	71	72	73	74	75	76
700	41	41.5	42	43	44	45

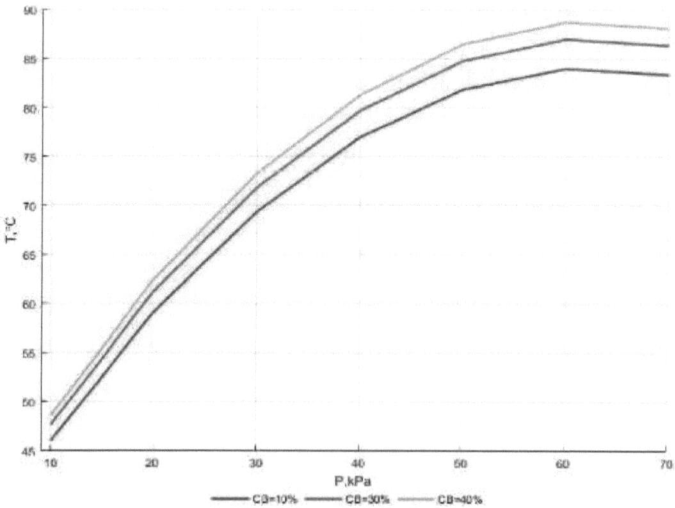

Fig. 2.16. *Dependence of the boiling point of liquorice root extract on the pressure in the apparatus at various dry matter contents of the solution*

CHAPTER 3

3. DETERMINATION OF BASIC OPERATING PARAMETERS OF THE EVAPORATION PROCESS OF LIQUORICE ROOT EXTRACT ROOT

3.1. Key requirements for the evaporation process

One of the key performance indicators of evaporators is the heat transfer coefficient. A stable value of the heat transfer coefficient determines the overall heat exchange effect, which characterises the sequential transfer of heat from the heating steam to the boiling solution. The value of the heat transfer coefficient is determined using the well-known formula

The phenomena occurring inside the boiling tubes of evaporators can be divided into two interrelated processes: heat transfer from the tube walls to the liquid and the movement of the liquid inside the tubes. The authors have studied the flow regimes of liquid in vertical boiling tubes. They are described by standard ideal displacement models, the defining parameters of which depend on the ratio of the pipes' geometric dimensions. The deformation of the hydrodynamic flow structure in a vertical pipe leads to a change in the spatial distribution of process parameters and heat transfer conditions along its height. The corresponding heat transfer zones exhibit different hydrodynamic flow structures within the vertical boiling tube, and each zone and each regime has its own specific heat transfer conditions. Furthermore, heat transfer under various flow regimes follows its own specific patterns and is analysed separately.

A review of the literature on heat transfer during liquid boiling in a vertical tube allows the following conclusion to be drawn. In the preheating and underheated boiling zones of the extract, the heat transfer coefficient depends mainly on the average temperature difference between the heating steam and the liquid, and on the liquid velocity in the tube.

Process intensification is achieved by inducing flow turbulence using various types of swirl generators.

In the bubble boiling zone, higher values are observed due to the disruption of the wall- d boundary layer of the liquid by growing and detaching solvent vapour bubbles. In this case, the intensification of the process is influenced by the size of the vapour bubbles and the frequency of their detachment, which depend on the properties of the liquid and the heat transfer surface material, as well as on the heat source power.

In the film boiling zone, the coefficient *K* has the highest value, which is caused by an increase in the evaporation rate of solvent molecules from the film and an increase in the mass flow rate of the film.

The film vaporisation zone is characterised by the breakdown or drying out of the liquid film. In this case, the heating surface is separated from the liquid film by a low-thermal-conductivity vapour layer, resulting in a sharp decrease in the value *of K*. In the work / / describes measures to prevent the exposure of boiling tubes.

The heat transfer coefficient *K* has a minimum value in sections of the vertical tube where dispersed liquid is present, since the thermal conductivity from the walls to the vapour phase is very low.

Very little research has been conducted into the distribution of the heat transfer coefficient along the height of a vertical pipe depending on the hydrodynamic structure of the flows. Meanwhile, this information is of great importance for the design of new apparatus for the evaporation of solutions from plant raw materials and for the optimisation of the process in existing evaporators.

The author of the work / shows the distribution of of the heat transfer coefficient as a function of the enthalpy of the secondary vapour. Analysis of this distribution reveals the stability of the heat transfer coefficient in the regions of bubble and film boiling of the liquid, and its sharp decrease upon the onset of vapour film boiling.

When calculating the bubble and film boiling regions of solutions in the tubes of evaporators, one can use average values of the heat transfer coefficient. This is because the distribution of the heat transfer coefficient along the height of a vertical tube has been quantitatively described for water and aqueous solutions. For solutions derived from plant materials, in particular liquorice root extract, the key characteristics of the evaporation process are not available. For evaporators to operate normally, the process must be conducted under bubble or film evaporation, which ensures minimal carryover of droplets of the product being evaporated with the secondary vapour.

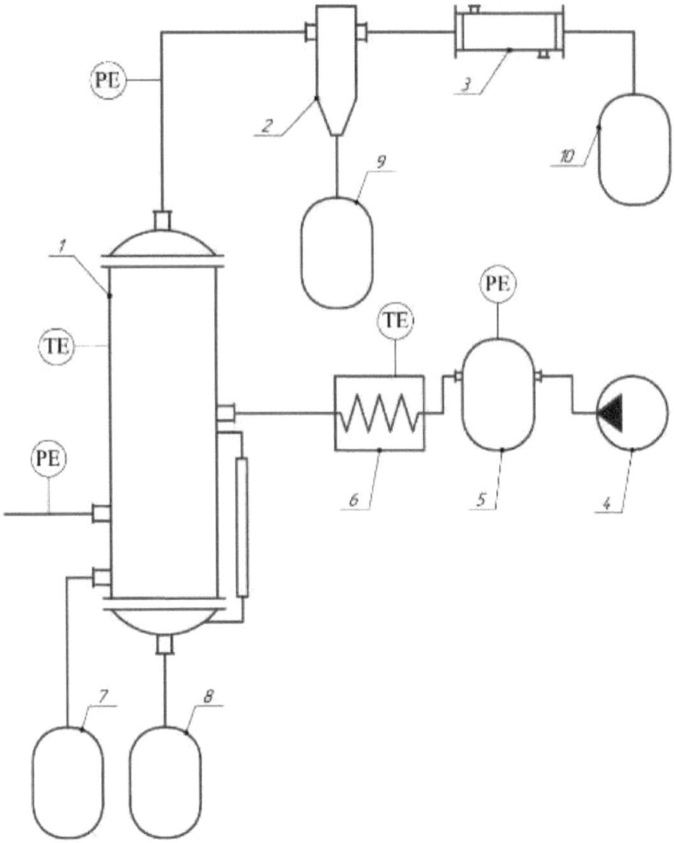

Fig. 3.1. *Diagram of the experimental setup*
1 – evaporator; 2 – trap; 3 – surface condenser;
4 – compressor; 5 – extract tank; 6 – thermostat;
7, 8, 9, 10 – vessels.

3.2. Description of the experimental setup

The experimental setup consists of an evaporator 1, a trap 2, a surface condenser 3, a tank 5, a compressor 4, a thermostat 6, measuring vessels 7, 8, 9, 10 and control and measuring instruments .

The initial extract, in the required quantity and at the appropriate concentration, is fed from vessel 5 into the evaporator. The extract is heated to a specific temperature in thermostat 6. The flow rate of the incoming extract is regulated by a reducing valve. Heating steam is supplied to the unit via a steam line through a throttling valve. The condensate from the heating steam is drained from the apparatus's heating chamber into a calibrated collection vessel. A pressure gauge is fitted to the steam line to monitor the steam pressure. Non-condensable gases, together with the juice vapour, are discharged into trap 2, where droplets of the carried-over product are separated and directed into collector 5, whilst the juice vapour is further directed into surface condenser 3. In condenser 3, the juice vapour condenses. The juice vapour condensate is directed into the collector. A thermometer was fitted to the body of the apparatus to measure the temperature of the juice vapour. The concentrated extract was discharged from the lower part of the apparatus into measuring vessel 6. The level of the extract, as well as its concentration within the apparatus, were maintained at constant values appropriate to the experimental conditions. The vacuum in the headspace of the apparatus was maintained by the volume of cooling water entering the condenser. The level of the extract in the evaporator was monitored using a sight glass.

3.3. Experimental methodology

The objective of the experiment on the experimental setup is to determine the heat transfer coefficients during the evaporation of liquorice root at various concentrations, to determine the carryover of droplets of the evaporated extract, and to determine the vapour pressure in the headspace of the evaporator.

The heat transfer coefficient was determined: by the amount of heating steam; by the amount of moisture evaporated from the extract during its concentration; and by the difference in the volume of solution entering and leaving the apparatus.

Sequence of experiments. An extract of a specific concentration was fed into the evaporator, and the required vacuum was maintained within the apparatus. The vacuum was created using a condenser. Each experiment lasted for 0.5– . The following process parameters were maintained during the experiments: initial and final solution concentration, heating steam temperature and solution boiling point, vacuum in the headspace, and the volume of heating and vapour condensate was measured.

To obtain reliable data on heat transfer coefficients, the experiments were conducted under steady-state thermal and process conditions in the apparatus. Heat transfer coefficients were determined by the following methods: a) from the heating steam condensate

$$K_1 = \frac{G_1 \cdot (r - t_{к.н.})}{F \cdot (T - t_{к.с})}; \quad \frac{Bm}{м^2 \cdot K}$$

б) from the secondary steam condensate

$$K_2 = \frac{G_2 \cdot (r_1 - t_{п.с.})}{F \cdot (T - t_{к.с})}; \quad \frac{Bm}{м^2 \cdot K}$$

в) based on the difference between the initial and final concentrations of the solution

$$\frac{G \cdot (CB_к - CB_н) \cdot (r - t_{п.с.})}{F \cdot CB - (T - t_{к.с})}; \quad \frac{Bm}{м^2 \cdot K}$$

To determine the values used in the calculation, the plant is equipped with appropriate measuring vessels and control and measuring instruments. The temperature of the incoming and outgoing juice, the heating steam condensate at the outlet of the apparatus, the temperature of the heating steam and the temperature of the juice vapour were measured using thermometers. The pressure of the heating and secondary steam was determined using pressure gauges. The concentration of the incoming and outgoing extract was determined using an RPL-2 laboratory refractometer. The quantity of extract entering the apparatus was measured using a tared container. The mass of the primary and secondary steam, as well as the quantity of extract leaving the apparatus, was determined using tared containers with an accuracy of 0.01 *kg.*

The experimental setup was also used to determine the vapour space pressure of the evaporator during the concentration of liquorice root extract. The vapour space pressure is the primary indicator characterising the operation of the apparatus, at which there is no carry-over of the product by the secondary vapour.

The volumetric vapour pressure was determined using the formula given by :

$$W = \frac{G_2}{V'' \cdot V_п}; \quad \frac{м^3}{м^3 \cdot ч} \qquad \textbf{(3.4)}$$

where: *O2* – the amount of secondary vapour condensate in *kg/h;*

V" – the specific weight of saturated secondary vapour in kg/m^3;

V_n - volume of the superheated space in m^3.

The maximum volumetric stress in the superheated space of the apparatus was determined by the onset of entrainment of extract droplets by the secondary steam. A dry matter content of 0.25% (by mass of the condensate) in the secondary steam condensate was taken as the reference value for the onset of entrainment. A qualitative check of the secondary vapour condensate for its dry matter content was carried out using a laboratory refractometer.

The determination of the dry matter content in the secondary steam condensate was carried out continuously. Samples were taken via a tap installed on the secondary steam condensate discharge line. The volume of secondary steam condensate obtained was determined using a calibrated container, as well as by the difference between the volume of extract entering the apparatus and the volume of extract leaving the apparatus.

1.4. Discussion of experimental results

The experiments were carried out under a vacuum corresponding to an extract boiling point of 85°C and with extract concentrations varying from 10% to 50% w/v. The evaporation temperature of 85°C is justified by the fact that at temperatures above 85°C, glyceric acid (the main indicator of product quality) decomposes during evaporation and the product takes on a dark colour, which does not meet GOST requirements.

Tables 5, 6 and 7 present experimental data obtained on the experimental setup. The experiments were conducted under steady-state conditions in triplicate over a period of 0.5 hours. Based on the experimental data, graphs were plotted showing the relationship between the vapour pressure and the heat transfer coefficient as a function of the dry matter content in the liquorice root extract.

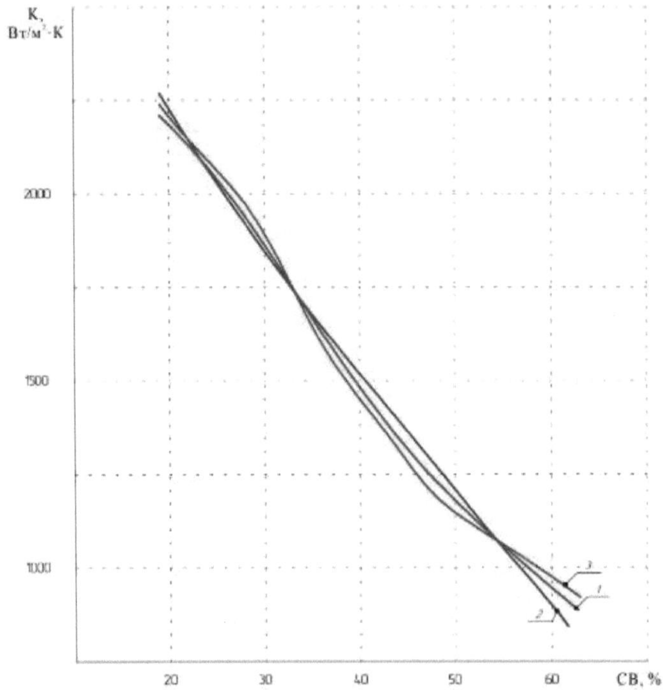

Fig. 3.2. *Dependence of the heat transfer coefficient on the concentration of the liquorice root extract: 1 – on the quantity of heating steam; 2 – on the amount of secondary steam; 3 – based on the volume of solution entering and leaving the evaporator*

Analysis of the curves (Figs. 3.1, 3.2) shows that the values of the heat transfer coefficients determined by the methods mentioned above differ little from one another. They can be represented by straight lines, the deviation between which is insignificant, and as can be seen from Table 5, the vapour pressure of the evaporator's vapour space decreases significantly with an increase in the concentration of the liquorice root extract.

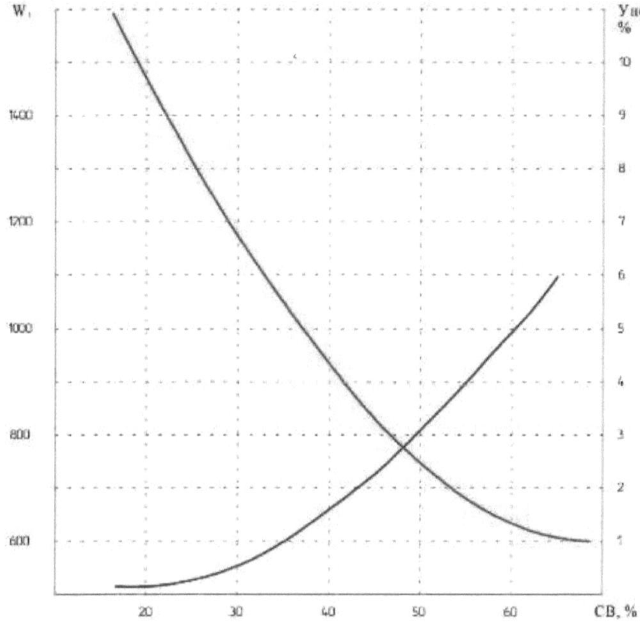

Fig. 3.3. Dependence of the critical vapour space pressure and product droplet entrainment on the dry matter content of the evaporated extract

At concentrations ranging from 10% to 40% dry matter in the extract, product entrainment with the secondary vapour ranges from 0.01% to 2.1%. At this level of droplet entrainment with the secondary vapour, any separator design completely separates the droplets from the secondary vapour. When evaporating liquorice root extract with a dry matter content exceeding 40%, the vapour pressure in the vapour space decreases significantly, and droplet entrainment by the secondary vapour increases to 3.4%. Such a level of entrainment makes it impossible to completely separate the droplets from the secondary vapour, regardless of the separator design of the evaporator. This leads to the conclusion that it is rational to carry out the evaporation of liquorice root extract in evaporators with a dry matter content not exceeding 40%.

CHAPTER 4

4. STUDY OF CONDITIONS WITHOUT A FOAMING REGIME FOR CONCENTRATING LIQUORICE ROOT EXTRACT

4.1. Boiling of foam-forming liquids

The process of evaporating extracts from plant raw materials is subject to complex patterns due to the influence of a large number of factors.

The evaporation of foaming solutions, such as liquorice root extract, is characterised by the great difficulty in separating product droplets from the secondary vapour.

A sufficient number of separators for evaporators have now been developed to ensure the proper separation of liquid and vapour after evaporation. However, when a vapour-liquid mixture consisting of foam (licorice root extract foam is very stable) enters the evaporator separator, the separation efficiency drops sharply and the liquid in the form of foam enters the condensation unit.

Our analyses have shown that persistent foaming in boiling tubes mainly occurs at the moments when boiling crises arise, particularly at the moment of the third crisis, as defined by A.A. Artikov.

The third crisis in vertical boiling tubes with an upward liquid flow is characterised by a low liquid flow rate during bubble boiling. Here, as the solution concentration increases, the steam flow rate increases and the intensity of bubble boiling rises.

However, there may come a point when the volume of liquid in the boiling tube is less than the volume required to sustain bubble boiling, yet the vapour velocity is still insufficient for a transition to the liquid-ring boiling regime. In this case, the vapour will entrain the liquid in the form of foam, i.e. intense foaming occurs. This regime was defined on the basis of the barbatage theory of bubbles in the boiling tube. With regard to the boiling of non-foaming solutions in the boiling tube, the following stages of boiling can be considered (Fig. 4.1): heating of the liquid, wall-adherent bubble boiling. If the wall is excessively heated, a vapour film will form, sharply reducing heat transfer – this is the first crisis. Sometimes a vapour breakthrough occurs, i.e. bubble boiling forms, which subsequently transitions to film boiling. Subsequently, when the liquid film dries out due to insufficient liquid flow, the heating surface is exposed – this constitutes the second boiling crisis in the tubes.

The boiling regime of solutions in tubes depends on a multitude of factors, among which the fluid flow rate and solution concentration play a major role.

According to the bubbling theory, it has been established that at certain flow rate ratios and when the liquid concentration is below the critical level, a point is reached where there is insufficient liquid to sustain film boiling; the boiling then transitions to bubbling boiling, and part of the liquid is carried away as vapour. For foaming liquids, a point is reached where the abundant foam generated moves upwards through the tube into the separation section of the apparatus. In the foaming section of the boiling tube, the heat transfer coefficient drops sharply.

During the third boiling crisis of the solution, its flow rate is determined by the equation:

$$G_{кр} = \rho_п \cdot n \cdot \sqrt{\frac{\pi^2}{32} \cdot \frac{\rho_ж - \rho_п}{\rho_ж} \cdot q \cdot d_{тр}^5} \qquad (4.1)$$

where ρ_v , ρ_l *are* the densities of the vapour and liquid, respectively, kg/m^3 ;

n — circulation ratio

d_{mp} - pipe diameter, *m*.

This theory can be used to explain the efficiency of solution circulation. Experimenters generally attribute the effect of solution circulation to an increase in the heat transfer coefficient due to increased turbulence in the liquid flow. Without denying the influence of this effect, attention may be drawn to another phenomenon: during liquid circulation, the flow rate and concentration of the solution entering the boiling tubes increase, thereby reducing the likelihood of a third boiling crisis occurring.

The study of the liquorice root extract concentration process was carried out on the experimental setup shown in Fig. 4.1 with the aim of determining the influence of flow rate and secondary vapour velocity on the separation of vapour-entrained droplets.

4.2. Description of the experimental setup

A diagram of the experimental setup is shown in Fig. 4.1. The setup consists of a boiling tube with an internal diameter d = 38 *mm* and a height H = 2700 *mm,* made of glass and vertically mounted on a stand *2.*

The setup includes vessel *6* for the extract, compressor *3* for feeding the extract from vessel *6* into the boiling tube *1,* regulator *4* for maintaining a constant extract flow rate, pressure gauge *5* for monitoring the pressure in vessel *6, and* valve 7 for regulating the extract flow rate. The experimental setup has an air supply system comprising a compressor *8,* a rotameter *10* for monitoring the air flow rate, separators *11* and *12* for separating the vapour-liquid mixture, made of glass and mounted vertically on stands *13* and *14,* a vessel *15* for the separated liquid, a vessel *16* for the condensate, and a vessel *17* for droplets carried away by the vapour stream.

Principle of operation of the experimental setup.

An extract of a specific concentration is continuously fed into the boiling tube. The compressor *6,* reducer 7, vessel 5 and valve *9* ensure a continuous and uniform supply of the extract. The supply of the specified air is provided by the compressor *10* and controlled by the rotameter *11* and valve *9.*

Inside the boiling tube, the liquorice root extract interacts with the air. The concentration of the liquorice root extract is determined using a refractometer. Once the tube has been filled with extract to a specified level, both the extract and air are fed into the tube simultaneously. The air containing droplets of extract enters separator *12,* where the product droplets are separated from the air. The separated extract is collected in the collector *20.* Liquid droplets not separated in the separator *12* are directed, together with the vapour, into separator *13.* The extract droplets separated in this separator enter vessel *21.* The unseparated product droplets are then directed into the collector *22* together with the air.

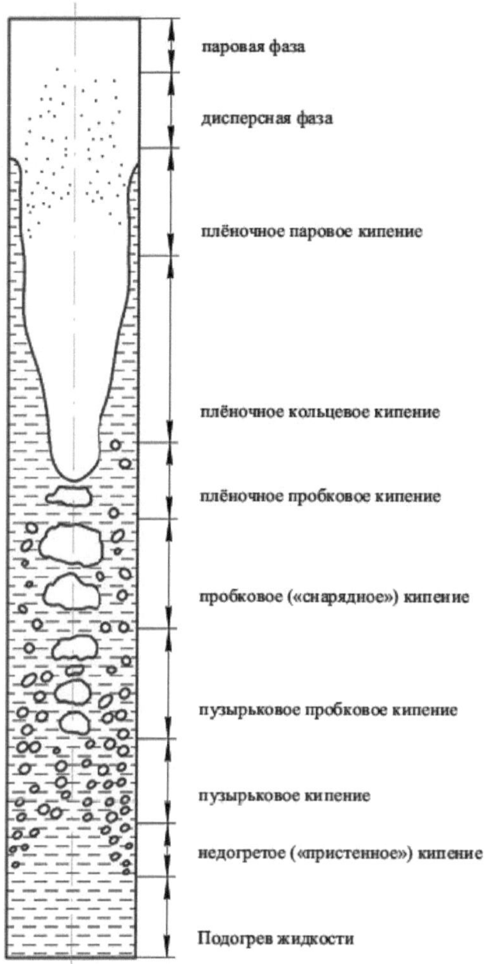

vapour phase
dispersed phase
film-type vapour boiling
film ring boiling
film plug boiling
plug (projectile) boiling
bubble plug boiling
bubble-cork boiling
underheated ('wall') boiling
Liquid preheating

Fig. 4.1 *Hydrodynamic structure of a two-phase flow during boiling of an extract in a vertical tube*

4.3. Experimental methodology at the experimental set-up

The objective of the experiments on the experimental setup is as follows:

- to determine the conditions for foam-free concentration of liquorice root extract:
- to determine the effect of the amount of secondary steam on the amount of product carried away by the secondary steam;

- to establish the optimal circulation regime for the extract, ensuring a foam-free evaporation process.

To ensure that the experimental conditions closely resembled real-world processes, liquorice root extract of varying concentrations was used.

The boiling tube is filled with the prepared extract so that approximately one-third of the tube remains empty.

The compressor is switched on, and the pressure regulator is used to set the required pressure in the extract vessel, ensuring the desired flow rate of the extract through the pipe. The pressure in the vessel is monitored by a pressure gauge. Using a valve, the flow rate is adjusted so that the level of the extract in the boiling tube remains constant. Once a steady-state flow of the extract has been achieved, the volumetric flow rate of the extract is determined.

When determining the extract flow rate, the extract is collected in a measuring vessel.

Air flow is monitored using a rotameter. During the experiments, air was used as the secondary vapour. In the experiments, the air (secondary vapour) and liquid flow rates are set such that a steady-state condition of uniform liquid boiling with minimal foaming and normal separation of the secondary vapour in the separator is achieved. Once the experimental setup has reached steady-state operation, the air velocity is determined using an anemometer. The air flow rate is recorded based on the rotameter readings.

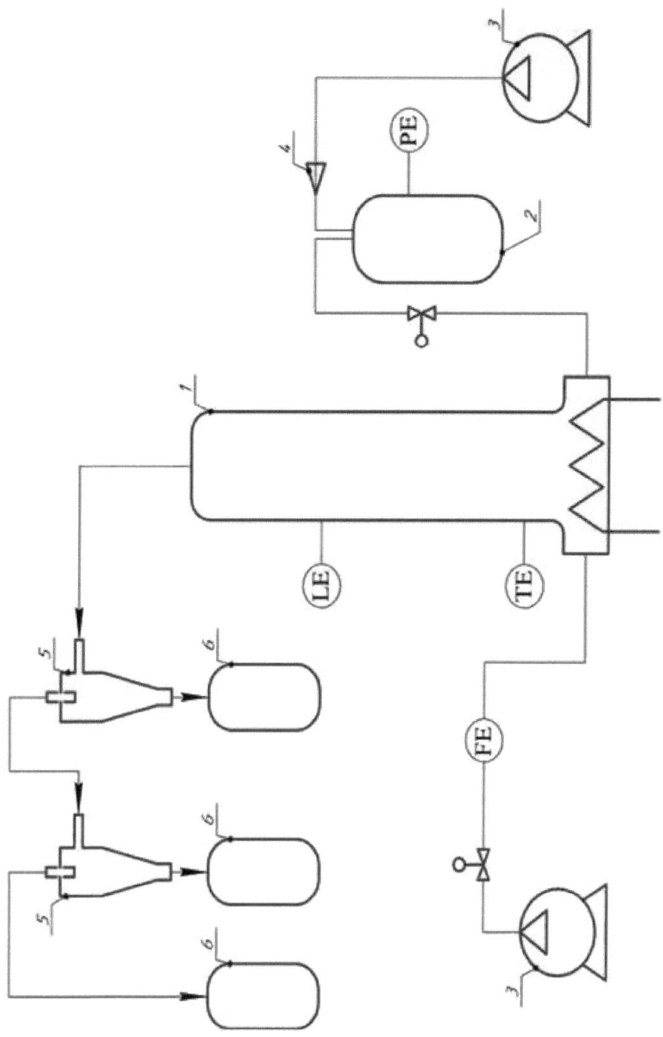

Fig. 4.2. Diagram of the experimental setup
1 – boiling tube; 2 – extract vessel; 3 – compressor;
4 – reducer; 5 – separator; 6 – vessel

4.4. Discussion of experimental results

The experiments were conducted at atmospheric pressure in the setup, varying the extract concentration from 10% to 40% and the air-to-extract flow rate ratio in the range from 0.3 to 1.2. The experimental results are given in Table

As can be seen, for the selected liquid flow rate, the maximum volume ratio of the vapour and liquid phases lies within the range of 0.5–1.0. Under these conditions, bubbling boiling occurs in the boiling tube without foam formation. A further increase in steam flow rate leads to the formation of a liquor from liquorice root extract, which impairs the separator's performance. Thus, for operating modes at specific liquid-to-steam ratios, the bubble boiling regime is maintained, provided that the aforementioned ratio of volumetric flow rates of steam (secondary steam) and liquid is observed:

$$\frac{V_п}{V_ж} = \frac{G_п}{G_ж} \cdot \frac{\rho_п}{\rho_ж} \qquad (4.2)$$

OR

$$\frac{G_п}{G_ж} = \left(\frac{V_п}{V_ж}\right) \cdot \frac{\rho_ж}{\rho_п}$$

Based on the results of our research, the mass ratios of vapour to liquid will be 0.5–1.0.

Thus, based on the experimental results, it has been established that for foam-free boiling of liquorice root extract in a tubular evaporator, the required ratio of secondary steam flow rate to the quantity of extract must be within the range of 0.5–1.0. If this ratio is maintained, the vapour-liquid mixture is separated almost completely; if a deviation from this ratio is allowed, the boiling tube becomes filled not with liquid but with foam, and a certain amount of product droplets is carried away by the secondary steam.

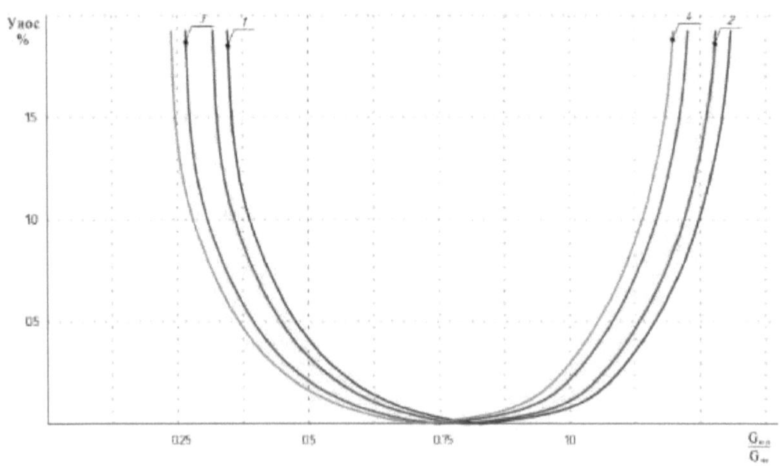

Fig. 4.3. *Dependence of the entrainment of product droplets by secondary steam on the ratio of the volumes of secondary steam and the extract being evaporated at various concentrations.*

1 – 40% SV; 2 – 30% SV; 3 – 20% SV; 4 – 10% SV

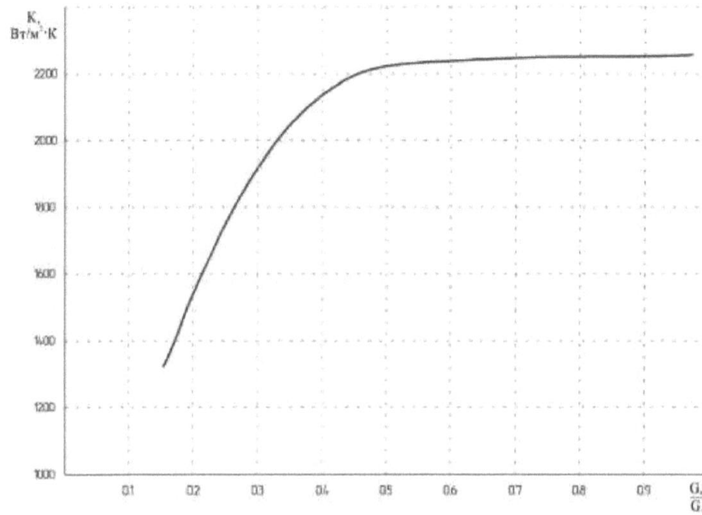

Fig. 4.4. *Dependence of the heat transfer coefficient on the ratio of secondary steam flow rate to evaporated solution flow rate.*

CHAPTER 5

5. STUDY OF A MATHEMATICAL MODEL OF THE STATICS OF THE EVAPORATION PROCESS OF LICORICE ROOT.

5.1. Fundamentals of mathematical modelling of concentration of extracts from plant raw materials

In recent years, the level of technical equipment in the food industry, particularly in extract production, has increased significantly. However, further progress in this direction is inconceivable without a detailed and in-depth study of food technology processes and equipment based on the methodological principles of systems analysis, which relies heavily on mathematical modelling (1,2,3,4–8,9,10,11).

A mathematical model of a process, being a system of mathematical descriptions of its elements from which the process itself is composed, must reflect the essence of the phenomena occurring within the system. A modelling algorithm must be defined for the mathematical model. The model must be considered in terms of its three aspects—semantic, analytical and computational—in conjunction with modern computing tools (12,13,14,15,16,17); methods of mathematical modelling allow, within a relatively short period of time and at relatively low material cost, the investigation of various options for the process's instrumentation, the study of its fundamental patterns, and the identification of potential for intensification.

Thermal processes play a major role in extract production; in particular, the heat-exchange process of liquorice root concentration constitutes the main technological step in extract production.

A systematic approach to the problem at hand plays a crucial role, necessitating the identification of cause-and-effect relationships that characterise the process flow. In this regard, it is advisable to examine and analyse the phenomena and effects of the concentration process across all structural elements of the apparatus. Such an analysis involves identifying the hierarchical structure of the process (as a typical physico-chemical system). When analysing and synthesising various options for process units for extract concentration, it is necessary to have the static equation that forms part of the mathematical model of the process.

When developing a mathematical model of the E.L.K. concentration process, the physico-chemical characteristics of the interacting phases, the design features and operating modes of the apparatus are taken into account, and certain assumptions are made.

The methodology for developing a mathematical model of the liquid-liquid separation evaporation process includes the following research stages: selection of the object, its study, justification of the object's structural diagram, formulation of a mathematical description of the phenomenon, derivation of a mathematical model of the process, selection of methods for solving the mathematical model of the object, and evaluation of the accuracy of calculations based on the proposed mathematical models of the evaporation process.

5.2. Mathematical model of the static behaviour of the evaporation process of liquorice root extract

For the optimal design of the process of concentrating liquorice root extract in vertical-tube evaporators, it is of great interest to study the distribution of parameters along the height of the boiling tubes in the evaporators.

The concentration of the liquorice root extract and its boiling point are taken as the determining parameters of the evaporation process. The distribution of these process parameters along the height of the apparatus's boiling tubes is studied by constructing a mathematical model of the static behaviour of the concentration process.

When constructing a mathematical model of the static behaviour of the evaporation process of extracts from plant raw materials, the main stage is the study of the characteristics of the processes occurring inside the boiling tubes of the apparatus.

To obtain a mathematical description of the process inside the boiling tubes, we select their elementary sections along the height, with initial and final boundaries. By drawing up the heat balance for this case, we obtain:

$$Q_{экj-1} + Q_{нрj-1} + Q_{гр} = Q_{экj} + Q_{нрj} \qquad (5.1)$$

where $Q_{экj-1} = G_{j-1} \cdot C_{j-1} \cdot t_{j-1}$ is the heat input from the extract being concentrated

extract;

$Q_{нрj-1} = G_{нрj-1} \cdot i_{j-1}$ – heat input from the solvent vapours;

$Q_{зм} = K \cdot \Delta\Gamma \cdot (t_{кн} - t_{j-1})$ -heat input from the heating steam via heat transfer through the walls of the boiling tubes;

$Q_{экj} = G_j \cdot C_j \cdot t_j$ heat loss with the concentrated extract;

$Q_{прjl} = G_{пр} \cdot i_{j-1}$ – heat loss via solvent vapours;

$Q = G_0 \cdot \frac{a_0}{a}$; $G_{пр} = G_0 \cdot \left(1 - \frac{a_0}{a}\right)$ - flow rates of the extract and solvent vapours;

$\Delta F = \pi \cdot d_{вн} \cdot n \cdot \Delta h$ -heating surface of the elementary sections.

The heat balance of an elementary section, when expanded, takes the following form:

$$G_{j-1} \cdot C_{j-1} \cdot t_{j-1} + G_{прj-1} + K \cdot \Delta F \cdot (t_{вн} - t_{j-1}) = G_j \cdot C_j \cdot t_j + G_{прj} \cdot i_j = 0 \qquad (5.2)$$

After substituting the values of the process design and technological parameters, we have:

$$G_0 \cdot \frac{a_0}{a_{j-1}} \cdot C_{j-1} \cdot t_{j-1} + G_0 \cdot \left(1 - \frac{a_0}{a_{j-1}}\right) \cdot i_{j-1} + K \cdot \pi \cdot d_{вн} \cdot n \cdot \Delta h \cdot (t_{кн} - t_{j-1}) - \\ - G_0 \cdot \frac{a_0}{a_j} \cdot C_j \cdot t_j - G_0 \cdot \left(1 - \frac{a_0}{a_j}\right) \cdot i_j = 0 \qquad (5.3)$$

After a series of mathematical transformations, we arrive at the following equation:

$$G_0 \cdot a_0 \cdot \left(\frac{i_j - C_j \cdot t_j}{a_j} - \frac{i_{j-1} - C_{j-1} \cdot t_{j-1}}{a_{j-1}} + \frac{i_j - i_{j-1}}{a_0}\right) = K \cdot \pi \cdot d_{вн} \cdot \Delta h \cdot (t_{вн} - t_{j-1}) \qquad (5.4)$$

Dividing both sides of the equation by JA, we obtain:

$$G_0 \cdot a_0 \cdot \frac{1}{\Delta h} \cdot \left(\frac{i_j - C_j \cdot t_j}{a_j} - \frac{i_{j-1} - C_{j-1} \cdot t_{j-1}}{a_{j-1}} + \frac{i_j - i_{j-1}}{a_0}\right) = K \cdot \pi \cdot d_{вн} \cdot (t_{кн} - t_{j-1}) \qquad (5.5)$$

Given that $a_j = a_{j-1} + \Delta a$ и $i_j = i_{j-1} + \Delta i$, we can write:

$$G_0 \cdot \frac{a_0}{\Delta h} \cdot \left(\frac{\Delta i - \Delta (C \cdot t)_j}{\Delta a} + \frac{\Delta i}{a_0}\right) = K \cdot \pi \cdot d_{вн} \cdot (t_{кн} - t_{j-1}); \\ G_0 \cdot a_0 \cdot \frac{\Delta}{\Delta h} \cdot \left(\frac{i - C \cdot t_j}{a} + \frac{i}{a_0}\right) = K \cdot \pi \cdot d_{вн} \cdot (t_{кн} - t_{j-1}) \qquad (5.6)$$

Moving from increments to differentials, we obtain:

$$G_0 \cdot a_0 \cdot \frac{d}{dh} \cdot \left(\frac{i - C \cdot t_j}{a} + \frac{i}{a_0}\right) = K \cdot \pi \cdot d_{вн} \cdot (t_{кн} - t_{j-1}) \qquad (5.7)$$

Equation (5.7) characterises the changes in process parameters with respect to the height of the boiling tubes of the evaporators, given their specified design and process parameters.

The boiling temperature of the extract t_{3KC}depends on its concentration and the pressure P in the apparatus and is determined from the regression equation obtained using the formula:

$$t_{кип} = f(a,P) \qquad (5.8)$$

The enthalpy of vapour of aqueous solutions depends on the boiling temperature of the extract and is determined by the formula:

$$i = 2482 + 1{,}97 \cdot t \text{ кДа/кг} \qquad (5.9)$$

The heat capacity of extract C is determined using the regression equation (2.14) as a function of its concentration and temperature.

$$C = f(a,t) \qquad (5.10)$$

The heat transfer coefficient is determined using the formula:

$$K = (2464 - 26{,}6) \cdot a \qquad (5.11)$$

To determine the condensation temperature of the heating vapour t_{KH}at its pressure $P_{(gk)}$,we use the equation:

$$t_{КН} = 86 + 0{,}15 \cdot P_{гк} \qquad (5.12)$$

Combining the above equations (5.8–5.12) into a system, we obtain the following mathematical model of the steady-state evaporation process of liquorice root extract in vertical tube apparatuses with distributed parameters:

I. $G_0 \cdot a_0 \cdot \frac{d}{dh} \cdot \left(\frac{i - C \cdot t_j}{a} + \frac{i}{a_0} \right) = K \cdot \pi \cdot d_{вн} \cdot (t_{кн} - t_{j-1});$

II. $t_{кип} = \frac{b_0 + b_1 \cdot CB - 0{,}014 \cdot p^2}{b_3 \cdot (b_4 - b_5 \cdot CB - b_6 \cdot CB^2)};$

III. $i = 2482 + 1{,}97 \cdot t;$

IV. $C = b_0 + b_1 \cdot CB + b_2 \cdot t + b_3 \cdot CB \cdot t + b_4 \cdot CB + b_5 \cdot t^2$

V. $t_{кн} = 86 + 0{,}15 \cdot P_{гк};$

VI. $K = (2464 - 26{,}6) \cdot CB.$

This process model makes it possible to study and analyse the process of concentrating liquorice root extract in the boiling tubes of the apparatus, as well as to identify the distribution of their parameters along the height of the boiling tubes. This makes it possible to design apparatus with optimal design parameters, as well as to identify the shortcomings and advantages of existing evaporator designs.

The optimal value of the main design parameter of evaporators—their heating surface area—can be calculated by formulating an optimisation problem and imposing constraints on the process parameters.

One method of calculation involves taking the specific productivity of the apparatus as the optimisation criterion, by which its optimal heating surface area is determined.

Given the specified process parameters G_o, A_o, *to* and design parameters *n*, $d_{(eH)}$, the required operating height of the apparatus's boiling tubes (*h*) is determined.

5.3 Algorithm for investigating the static behaviour of the evaporation process of liquorice root extract on a computer

The static behaviour of the liquorice root extract concentration process, based on its mathematical model (5.7), is investigated using a computer according to the following algorithm:

1. Numerical values are set for the parameters: the diameter of the tubes d_{eH} and their number *n*, the concentration, temperature and flow rate of the weak extract entering the apparatus – G_o , and the pressure of the heating steam in the apparatus casing P_{gk} and *K*.
2. From equation (5.12), which characterises the steam pressure in the apparatus's heating chamber P_{gk}, the condensation temperature of the heating steam t_{KH} is determined.
3. The increment in the concentration of the extract, Δa , is specified, bearing in mind that $\Delta a = a_j - a_{j-1}$.
4. The boiling temperature of the extract is determined from the regression equation (5.8) obtained from experimental data.

The heat transfer coefficient is determined from equation (5.11).

5. Substituting the value *of* t_{3KC} into equation (5.8) determines the enthalpy of the solvent vapours.
6. By substituting the values of the parameters *a* and *t* into equation (5.10), the heat capacity of the extract *C* is determined.
7. By substituting the numerical values of the parameters t_{KH}, *t*, *i* and *C* into equation (5.7), the concentration and temperature at this section of the pipes are calculated.
8. For given values *of P* and *K*, the calculation is repeated from step 3 to step 7, followed by a check that the condition $a = 40\%$ and $t = 85$ °*C* is satisfied.

The calculation is carried out until the optimal extract concentration $a = a\_(\text{pt})$ is reached at a given pressure $P = P\ (t = 85)$, ensuring optimal processing conditions for the extract.

Based on this algorithm, a flowchart has been drawn up to study the evaporation process using its mathematical model with the aid of a computer (Fig. 5.1), along with a program for its implementation.

The flowchart of this algorithm, drawn up for the computer-based study of the liquorice root extract concentration process based on its mathematical model, is shown in Fig. 5.1. The program for its implementation is provided in the appendix.

5.4 Investigation of the steady-state behaviour of the liquorice root extract concentration process

The aim of the study was to determine the optimal dimensions of the evaporator design, namely *h and F,* for a given specific capacity. To this end, we investigated the distribution of process parameters (*t* and *C)* along the height of the boiling tube.

It is worth noting here that the heating surface of the apparatus depends on the height of the boiling tubes:

$$F = \pi \cdot d_{вн} \cdot n \cdot h$$

We therefore consider that the distribution of process parameters along the height of the boiling tubes is identical to their distribution along the height of the apparatus.

To investigate the steady-state behaviour of the liquorice root extract concentration process based on its mathematical model, a programme was written in the BASIC programming language to implement the task on a computer.

The studies were conducted at an initial concentration of the weak extract entering the boiling tube of the evaporator, $a = 10$ %, with a flow rate of $G_o = 1290$ *kg/h.* The condensation temperature of the heating steam in the apparatus jacket is $t_{KH} = 140$ *°C* at its design pressure $P_{chk} = 360$ *kPa.* The design absolute pressure, according to calculations, is within the range $P_j = 59$ *kPa.*

The numerical value of the heat transfer coefficient in the apparatus, based on experimental data, is taken within a wide range from 0.5 to 5 $W/m^2 \cdot °C$.

The increase in extract concentration in the calculations is taken as $a = 2\%$.

Based on the results of the study of the process using its mathematical model, distribution curves for the extract concentration and temperature along the height of the boiling tubes of the evaporator were constructed.

The study of the mathematical model of the static behaviour of the liquorice root extract concentration process was carried out by determining the distribution of extract concentration along the height of the heating tubes of the heating surface. Given the number of boiling tubes *n* and their diameter *d*, the distribution of concentration and temperature along the height of the boiling tube can be examined. Studies were carried out for various values of the heat transfer coefficient. As can be seen from the figure, within the specified concentration range of the liquorice root extract (t = 85 °C), the concentration distribution is parabolic in nature. The degree of concentration is greatly influenced by changes in the average value of the heat transfer coefficient. A change in the heat transfer coefficient along the height of the boiling tube mainly contributes to a shift (weakening) of the initial boiling zone, and thereafter the concentration curve runs parallel to the concentration line at the average value of the heat transfer coefficient. In our studies, we mainly considered cases where there is a gradual increase in the heat transfer coefficient and cases with a pronounced maximum value of the heat transfer coefficient in the middle section of the boiling tube.

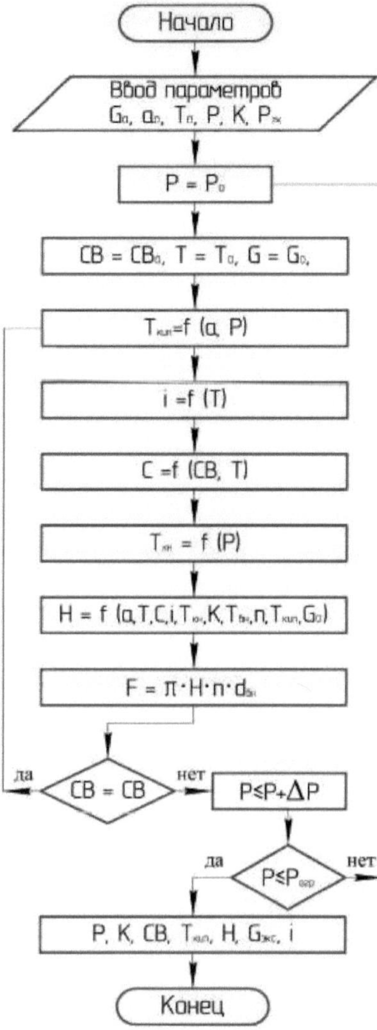

Fig. 5.1. *Block diagram of the algorithm for calculating the mathematical model of the static behaviour of the liquorice root extract concentration process on a computer*

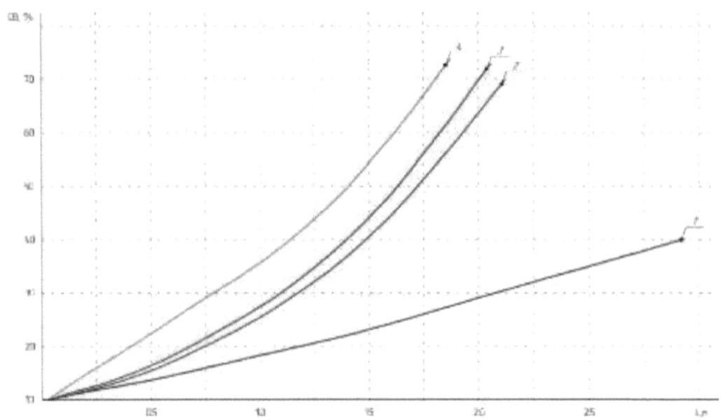

a)

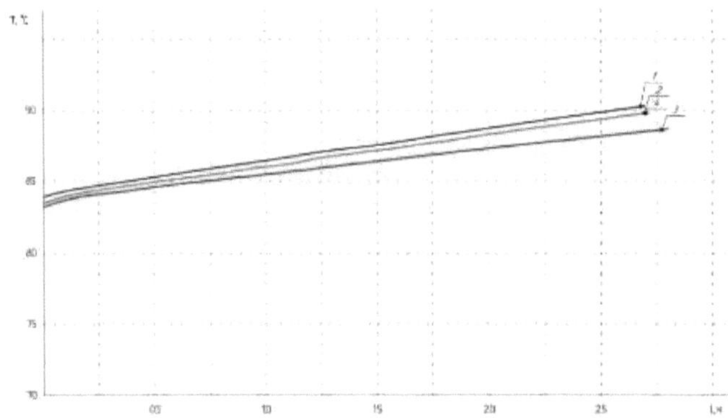

b)

Fig. 5.2. *Distribution of the parameters of the liquorice root extract concentration process along the height of the boiling tubes of the evaporator*

1 – at K = 500 W/m^2–K; 2 – at K = 1000 W/m^2–K; 3 – at the average value of K; 4 – at K = 1500 W/m^2–K;

In the example under consideration, the difference between the heating surfaces for the case of liquorice root concentration with a variable heat transfer coefficient and its average value is no more than 15%. Therefore, for engineering calculations, the method of determining the heating surface using the average heat transfer coefficient may be used. To this end, to determine the concentration distribution along the height of the boiling tube for a given outlet concentration, the corresponding height of the boiling tube is selected . For example, if the calculated heat transfer coefficient is 1500 *W/m·K,* the number of tubes is 84, the diameter d = 0.38 *mm*, and the distribution curve for concentrating liquorice root extract from *10%* to *40%* requires a boiling tube height H = 1.25 *m,* taking into account the safety factor (let us assume 1.5), the recommended height of the boiling tube will be 2.5 *m.*

CONCLUSION

This paper presents the results of experimental studies aimed at improving the efficiency of the liquorice root extract evaporation process by intensifying the development of optimal operating conditions based on a comprehensive set of investigations.

The thermophysical characteristics of liquorice root extract as a function of its concentration and temperature have been determined for the first time. Patterns have been established and empirical relationships derived for calculating the values of the thermal conductivity, heat capacity, thermal diffusivity and density of liquorice root extract within the investigated range of temperatures and concentrations.

The rheological properties of liquorice root extract have been investigated. The flow patterns of pseudoplastic fluids, to which liquorice root extract belongs, have been established; an equation characterising the flow of the extract has been selected and the coefficients of the equation have been determined.

The boiling point of the liquorice root extract was determined at various dry matter contents and pressures in the evaporator. An empirical relationship was obtained for calculating the boiling point of the extract within the studied range of concentrations and pressures in the apparatus.

The main parameters characterising the operation of evaporators have been determined - the heat transfer coefficient for liquorice root extract of varying concentrations at the maximum temperatures of the heating steam and the boiling point of the solution, and the maximum vapour pressure in the headspace of evaporators with natural circulation during the evaporation of foaming solutions, ensuring the evaporation process without entrainment of droplets with the secondary steam.

The influence of secondary steam flow rate and velocity on the removal of secondary steam without droplets of the evaporated product has been investigated with the aim of establishing the optimal ratio of the liquid phase of the extract to the amount of evaporated solvent, ensuring a droplet-free evaporation process.

A mathematical model of the process statics has been developed, and an algorithm and computer program modules have been proposed for calculating the concentration process of liquorice root extract. Based on the research results, distribution curves of process parameters have been plotted against the height of the boiling tubes – the main element of evaporation concentration apparatus. Based on an analysis of the results of the study of the process dynamics, the optimal design parameters of the evaporator have been determined.

REFERENCES

1. http://strategv,regulation, gov.uz/uz/document-'l
2. Musaeva N.X., Qodirova R.X. Research and experimental determination of thermophysical properties of highly foaming solutions// International Journal of Psychosocial Rehabilitation. England 2020. pp. 4611–4620.
3. Study of a mathematical model of the static process of evaporating liquorice root extract// Universum: Chemistry and Biology. Scientific journal. Moscow 2019.
4. Research on the mathematical model of the statistics of the steaming process of tomato paste.// International Journal of Advanced Research in Science, Engineering and Technology. 2019
5. O. Guclu-Ustundag, F. Temelli. Proceedings of the 5th International Symposium on Supercritical Fluids, Atlanta, Georgia, USA, 8–12 April 2000.
6. N.T. Machado, G. Brunner. Proceedings of the 5th International Symposium on Supercritical Fluids, Atlanta, Georgia, USA, 8–12 April 2000.
7. A.L. Chassagnez-Mendez, N.T. Machado et al. Proceedings of the 5th International Symposium on Supercritical Fluids, Atlanta, Georgia, USA, 8–12 April 2000.
8. F. Gironi, M. Maschetti. Chem. Eng. Sci., vol. 61, No. 15, pp. 5114–5124, 2006.
9. M. Herrero, A. Cifuentes, E. Ibáñez. Food Chemistry, vol. 98, No. 1, pp. 136–148, 2006.
10. Artikov A.A. Optimisation of the preliminary distillation process of cotton micelles. - Thesis ... Candidate of Technical Sciences. - TASHPI, Tashkent, 1970.
11. Artikov A.A. Processes and apparatus in food production. Mathematical modelling. Heat transfer processes. Evaporation. – Tashkent, Ukituvchi, 1983. – 123 pp.
12. Artikov A.A., Mamatkulov A.Kh., Yakhshimuradova N.K., Dodaev K. System analysis of solution concentration using an inert gas. – Tashkent, Fan, 1987. – 164 pp.
13. Artikov A.A. Enthalpy of secondary vapour and classification of evaporators. Proceedings of Universities, Food Industry, 1982, No. 2, pp.

83–86.
14. Artikov A.A., Khudayberdiev A.A., Nosyrova 3. Optimisation of processes in the oil and fat industry. Abstracts for the All-Union Seminar. – Krasnodar, KPP, 1983. – pp. 43–45.
15. Anoshin I.M. Theoretical foundations of mass transfer processes in food production. – Moscow: Food Industry, 1970. – 344 pp.
16. Asmaev M.L., Kornilov Yu.G. Modelling of food production processes. - Moscow: Light and Food Industry, 1982. - 177 pp.
17. Askerov N.R. Thermophysical Properties of Grape Products from Azerbaijan. - PhD thesis. Azerbaijan Institute of Oil and Chemistry, Baku, 1988.
18. Alekseev E.L., Pakhomov V.F. Modelling and optimisation of technological processes in the food industry. Moscow: Agropromizdat, 1987. – 271 pp.
19. Belik V.G., Kostanzhi I.L. Reference Guide on the Modelling and Optimisation of Heat Exchange Equipment in the Sugar Industry. - Moscow: Agropromizdat, 1986. - 270 pp.
20. Belkin I.M., Vinogradov G.V., Leonov A.I. Rotating Machinery. – Moscow: Mashinostroenie, 1968. – 272 pp.
21. Boyarinov A.I., Kafarov V.V. Optimisation Methods in Chemical Technology. - Moscow: Khimiya, 1969. - 564 pp.
22. Belik F.G., Fedotkin I.M. Modelling and optimisation of evaporators with steam extraction. - Kyiv: Tekhnika, 1974. - 130 pp.
23. Beloborodov V.V. Main processes in vegetable oil production. - Moscow: Pishchepromizdat, 1966. - 478 pp.
24. Bereznyak E.D. Issues in the intensification of heat exchange processes in the production of concentrated tomato products. Author's thesis for the degree of Candidate of Technical Sciences. Moscow, 1980.
25. Bolshakova N.V. Study of the process of liquid evaporation under conditions of radiative heat transfer. Author's thesis for the degree of Candidate of Technical Sciences. Moscow, 1971.
26. Beloborodov V.V. Main processes in vegetable oil production. Moscow: Pishchepromizdat, 1966. 448 pp.
27. Viter A.V., Marchenko A.N., Marchenko P.N., Shanovich O.P. Application of a vacuum evaporation unit with a centrifugal evaporator for the concentration of plant extracts. - Chem. Pharm. J., 1975, 9, No. 2. - pp.

47-50.
28. Viesturs U.E. Research and improvement of methods and apparatus for aerobic fermentation: Dissertation ... Candidate of Technical Sciences. – Riga, 1967. – 203 pp.
29. Viesturs U.E., Kristapsons M.J., Dolgitser N.S. Methods and devices for foam suppression in microbiological processes. Moscow: 1973. – 127 pp. /Series U1 'Microbiological Industry'. Information review / Glavmikrobioprom, ONTITEI Mikrobioprom /.

MIX
Papier aus verantwortungsvollen Quellen
Paper from responsible sources
FSC® C105338

Printed by Books on Demand GmbH, Norderstedt / Germany